Joseph Stephen Cullinan

Joseph Stephen Cullinan in 1903

JOSEPH STEPHEN CULLINAN

A Study of Leadership
in the Texas Petroleum
Industry, 1897–1937

JOHN O. KING

Published for
The Texas Gulf Coast
Historical Association by

Vanderbilt University 〰 Press
Nashville, 1970

International Standard Book Number 0–8265–1155–4
Library of Congress Catalogue Card Number 70–112935

Printed in the United States of America by
Heritage Printers, Inc., Charlotte, North Carolina
Bound by Kingsport Press, Kingsport, Tennessee

Contents

Illustrations

Preface

THE TEXAS petroleum industry began at Corsicana late in the nineteenth century and within a decade after the Spindletop discovery of 1901 it had reached a level of development profoundly changing the structure of the entire American industry. This rapid growth, in an essentially agrarian state hundreds of miles from pre-existing petroleum development, offered abundant opportunities to experienced oilmen willing to migrate into Texas and lead this new industry.

The subject of this study was one of those accepting this challenge. Joseph Stephen Cullinan was an aggressive Pennsylvania oilman whose ability and industriousness had enabled him to rise from an oil field laborer to executive rank within the Standard Oil organization. Thus, in 1897, he brought to the embryonic Texas petroleum industry vigorous managerial direction, contemporary technological practices, and access to eastern investment capital.

But the transition to an industrial order is rarely accomplished without turmoil and controversy, and the impact of the petroleum industry upon Texas was no exception. Cullinan, a leader from the beginning, was highly successful in achieving the industrial stability necessary for continued economic growth. Yet he also experienced disappointments and failures. His business plans sometimes had to be changed when a conflict developed with the state's prevailing political attitudes and legal authority. His highly individualized style of managerial direction eventually brought internal dissension within the major petroleum company that he organized and led to corporate maturity. This study examines Cullinan's career, to gain a fuller understanding of growth and change within the Texas petroleum industry during these crucial years of its development.

Houston, Texas   John O. King
October 1969

Acknowledgments

THE AUTHOR wishes to express his appreciation to those who offered assistance in this study. His thanks go particularly to Miss Nina J. Cullinan for permission to examine her father's records, the Joseph Stephen Cullinan Papers, deposited in the Texas Gulf Coast Historical Association Archives, M. D. Anderson Library, University of Houston; and to Mrs. Edward W. Kelley for permission to use her father's papers, the James Lockhart Autry Papers, deposited in the W. W. Fondren Library, William Marsh Rice University. The cooperation and helpfulness of Kenneth E. McCullam, historian of the Texas Company, New York City, in allowing the author to examine his company's archives are also greatly appreciated.

Several personal conferences with the late dean of Texas Oil historians, Charles A. Warner, were invaluable. The comments and suggestions offered by James H. Durbin, former business associate of J. S. Cullinan, were essential to the completion of this work. Likewise, several interviews with James A. Clark, oil historian and journalist, were also very rewarding. The author is further indebted to Professor James A. Tinsley, Executive Director of the Texas Gulf Coast Historical Association, for permission to use that organization's petroleum collection, particularly the extensive Charles A. Warner collection.

The patient help and guidance given by Professor Dewey W. Grantham of the Department of History, Vanderbilt University, in directing this study through its initial stages as a doctoral dissertation, is gratefully acknowledged. Lastly, the author wishes to thank his wife, Betty, whose help and encouragement made this work largely possible.

Joseph Stephen Cullinan

1. The Corsicana Oil Field: Early Development and Disorder, 1894-1897

ON OCTOBER 15, 1897, a delegation of worried businessmen from Corsicana, Texas, traveled the sixty miles north to Dallas. Ironically, their worries stemmed from the accidental discovery of oil at Corsicana in 1894—far from an unmixed blessing. The three-year petroleum development which stemmed from the discovery had been largely the work of these same businessmen, and it had brought perplexing and frustrating problems which now threatened the town's economic future. There was a desperate need for further capital to expand the oil field's full potential and, more important, there was a growing realization that the complexities of the petroleum industry demanded experienced managerial direction. But adequate capital and experienced managerial talent could not be found in Texas; Corsicana businessmen finally were forced to seek this all-inclusive help from outside the state, from "foreign" sources.

The journey to Dallas was an attempt to find a solution for Corsicana's problems. The town's mayor, James E. Whiteselle, previously had written an experienced oilman, Joseph S. Cullinan, of Washington, Pennsylvania, who agreed to interrupt a business trip to California for a brief inspection of the Texas petroleum development. The Corsicana group was on its way to meet him, determined to impress the visitor with Texas hospitality. In addition to meeting him in Dallas, the delegation had arranged a tour of the recently-opened Texas State Fair and they had scheduled a meeting with Governor Charles A. Culberson, also in Dallas for the opening of the fair.[1]

The Pennsylvania oilman was undoubtedly impressed and flattered by these social amenities, but he was restless and eager to begin the short rail journey south to Corsicana with his hosts a day later and to bring his trained eye upon the field. The brief inspection tour soon lengthened into a thorough three weeks' examination, and the visitor's California plans were canceled: Joseph S. Cullinan had begun his long career within the dynamic and challenging structure of the Texas petroleum industry.

Born December 31, 1860, near Sharon, Mercer County, Pennsylvania, Joseph Stephen Cullinan was the oldest son, second of the eight children[2] of John Francis and Mary Considine Cullinan.

1. Dallas *Morning News,* October 16, 17, 1897.
2. The other children of this marriage were Margaret Ann (Mrs. Thomas A. Hughen), born February 3, 1858, Mercer County, Pennsylvania, died January 11, 1908, Washington, Pennsylvania; Michael Patrick, born January 19, 1865, Venango County, Pennsylvania, died January 20, 1927, Laredo, Texas; Mary Ellen, born July 1, 1868, Venango County, Pennsylvania, died September 1, 1941, Washington, Pennsylvania; Anna, born November 29, 1870, Venango County, Pennsylvania, died August 1, 1956, De Land, Florida; Catherine, born April 20, 1873, Venango County, Pennsylvania, died June 7, 1930, Washington, Pennsylvania; Jane (Mrs. Richard F. Sawyer), born April 1, 1875, Venango County, Pennsylvania, died December 18, 1961, De Land, Florida; John Francis, born August 17, 1877, Venango County, Pennsylvania, died July 15, 1955, Alexandria, Minnesota.
Information on the background of the Cullinan family and on the education and early career of J. S. Cullinan came from the following sources: "Cullinan Family Genealogy" folder, Joseph Stephen Cullinan Papers, Texas Gulf Coast Historical Association Archives, University of Houston Library, Houston, Texas (hereinafter cited as Cullinan Papers); James A. Connelly (Sharon, Pennsylvania) to author, June 4,

His parents, both natives of County Clare, Ireland, had immigrated with their separate families to the United States in the 1840s. They had been married at Dubuque, Iowa, April 13, 1856, and shortly afterward they had moved to western Pennsylvania, settling on a small farm outside Sharon, at a rural community appropriately called "the Irish Settlement."

Young Joe Cullinan attended public school through the lower grades and found employment at the age of fourteen in the nearby oil fields of western Pennsylvania. The following years spent in this rugged area as messenger boy, oil-wagon teamster, pipeline laborer, and drilling crew member brought Cullinan a wealth of basic vocational experience. In 1882, he became an employee of Standard Oil's major transportation affiliate, the National Transit Company of Oil City, Pennsylvania. He soon advanced to the job of foreman of pipeline and storage construction crews as the company extended its facilities into new fields in southwestern Pennsylvania and West Virginia.

Young Cullinan's work with the National Transit Company brought him to the attention of such key executives of the Standard Oil group as Calvin N. Payne, Daniel O'Day, and Henry C. Folger, Jr. On the recommendation of Payne,[3] general manager of National Transit, Cullinan was transferred in 1888 to Standard's major transportation affiliate in the Ohio-Indiana oil field, the Buckeye Pipe Line Company, where he was appointed superintendent of that company's natural gas and tankage departments with headquarters at Lima, Ohio. This assignment was, eventually, to bring a felicitous change in the young bachelor's personal life, as well: on April 14, 1891, Joe Cullinan married Miss Lucy Halm, daughter of a Lima merchant.

By 1893, Cullinan was obviously marked for continued advancement in his company. He had been made a division superintendent

1963; James H. Durbin (New York) to author, July 8, 1961; "Employee Character Sketches" folder, the Texas Company Archives, (New York, N. Y.).

3. On the role of National Transit Company within the Standard Oil structure and the early association of Calvin N. Payne with Standard Oil, see Ralph W. Hidy and Muriel E. Hidy, *History of Standard Oil Company (New Jersey): Pioneering in Big Business, 1882–1911* (New York: Harper and Brothers, 1955), pp. 22, 158, 172.

at Washington, Pennsylvania, for another Standard Oil subsidiary, the Southwest Pennsylvania Pipe Line Company. But in 1895, he left his long-time employers: eager to try his hand at independent management, he organized an oil equipment manufacturing concern, the Petroleum Iron Works.

The firm was to specialize in fabricating and erecting steel storage tanks and steam boilers. It was organized as a partnership, with Cullinan as the managing partner, assisted by active partners Charles H. Todd and Edward G. Wright, both of Washington, Pennsylvania. Additional capital was furnished by silent partners, John Slater and W. C. McBride, also of Washington, and Joseph P. Sweeney and Edward A. Ryan, of Sistersville, West Virginia. All of these participants had wide contacts in the nearby southwestern Pennsylvania-West Virginia oil field development. Wright, McBride, and Sweeney, like Cullinan, were former employees of Standard Oil companies. Joseph P. Sweeney, particularly, must have furnished some help to the new firm, for he had been sent to Sistersville in 1892 as local agent for the Joseph Seep Purchasing Agency, Standard Oil's major crude oil-buying affiliate. Settling permanently in Sistersville, Sweeney was elected city recorder in March 1896. He was re-elected, the following year, "by the largest vote ever given a candidate in the city's history."[4]

Although no financial records of this early partnership are extant, the Petroleum Iron Works was unquestionably profitable, despite heavy competition from a proliferation of tankage firms servicing the Pennsylvania-West Virginia petroleum area.

A full-page advertisement which appeared in 1898 in *The Derrick's Hand-Book of Petroleum*, the country's major oil-trade publication, carried a photograph of the Petroleum Iron Works, showing its substantial factory and office buildings in Washington, Pennsylvania. The advertisement noted that Cullinan, Wright, and Todd were the active partners and proclaimed that the firm made a "Specialty of Pipe Line and Refinery Work."[5] Further

4. *The Derrick's Hand-Book of Petroleum: A Complete Chronological and Statistical Review of Petroleum Developments from 1859 to 1899*, 2 vols. (Oil City, Pa., 1898–1900), I, 700.
5. *Ibid.*, lxvi.

examination of the advertisement section of this publication for 1898 reveals the highly competitive nature of the storage-tank fabrication and construction business: twelve such firms, located in the Western Pennsylvania-West Virginia field, advertised in one volume alone.

Cullinan, as managing partner of his company, indicated immediate prosperity by purchasing a modest home in Washington, Pennsylvania, and maintaining his growing family in comfortable circumstances.[6]

Further evidence of the worth of the Petroleum Iron Works came later. In 1899, even after Cullinan had transferred his energies to Texas, he purchased the interests of his Pennsylvania and West Virginia partners and reorganized the concern as a corporation under the same name. In the new firm, Cullinan was the majority stockholder, while his former partners, Wright and Todd, managed the company and held minority stock interests. In 1909, the firm moved its headquarters from Washington to a larger plant site at Sharon, Pennsylvania.[7] The company grew in value as Cullinan's Texas oil interests expanded for, as would be expected, Petroleum Iron Works furnished most of the tankage and pipeline equipment used by these ventures.[8]

His early years as managing partner of the Petroleum Iron Works Company were vital in shaping Cullinan's career. He gained further experience and self-confidence from the exercise of independent managerial authority. In extensive travel for his firm through the Appalachian and Middle West petroleum regions, he made further business contacts and often renewed older friendships among the oil fraternity. In the design and adaptation of equipment offered

6. Four of the five children of Joseph S. and Lucy H. Cullinan were born at Washington, Pennsylvania: John Halm (1893), Craig Francis (1894), Nina Jane (1896), and Margaret Ann (1898). The fifth child, Mary Catherine, was born at Corsicana, Texas (1901).

7. Application for listing on the New York Stock Exchange, American Republics Corporation, November 5, 1923, Cullinan Papers.

8. The *Oil Investors' Journal* for September 5, 1907 (p. 3), for example, says "The Petroleum Iron Works Company . . . has obtained a contract to build six additional 37,500-barrel steel tanks for The Texas Company on the Clinton tank farm south of Tulsa." J. S. Cullinan was, of course, president of the Texas Company from 1902 until 1913.

for sale by his firm, Cullinan gained firsthand knowledge of the problems and practices of different petroleum-producing areas. Above all, he learned to shoulder the challenge and responsibility of on-the-spot decisions often demanded in highly competitive business situations.

As the Texans who met Cullinan at Dallas immediately perceived, his physical presence also invited—actually, demanded—respect and confidence. He stood just under six feet tall, with a muscular body hardened by years of vigorous oil-field activity. It was during these early years in the field that he acquired the nickname "Buckskin Joe" from his Pennsylvania contemporaries, who saw similarities in Cullinan's aggressive, abrasive qualities as a crewboss determined to let nothing interfere with job completion and the rough leather used for oil-field work gloves and shoes. This nickname, of course, was then unknown to his Corsicana hosts; the Texans were to know a more mature Cullinan with a high degree of executive polish gained from his work at Standard Oil. In succeeding years, on rare occasions, a particularly slothful employee or associate would feel, in verbal discipline from Cullinan, a latent sting reminiscent of the "buckskin" days.[9] And the sobriquet of earlier days continued to symbolize some of Cullinan's business qualities. Texans, from 1897 on, were thus to meet an experienced, aggressive Pennsylvania oilman in his physical prime, emboldened by the first blush of managerial success and with a restless eye for further opportunity.

While it was the Corsicana field that drew him to Texas, Cullinan and his colleagues of the eastern oil industry knew of the presence of oil and gas deposits within the Lone Star State long before the Corsicana discovery. There were seepages of oil and gas in Nacogdoches, Hardin, and Angelina counties before the Civil War and, in 1866, a Confederate Army veteran, Lyne T. Barrett, drilled the state's first successful oil well at Oil Springs, Nacogdoches County. This well at first produced ten barrels of crude oil daily, but the flow soon diminished and further exploration was abandoned. In the 1870s, drilling ventures at Sour Lake and

9. "Employee Character Sketches" folder, Texaco Company Archives.

Saratoga, in Hardin County, also produced oil, but not in sufficient quantities to encourage commercial development. During the same decade, the presence of oil and gas was noted at several other points in Texas; but these discoveries were often the accidental result of water-well drilling or construction projects. By 1880, evidences of petroleum deposits had been recorded in eighteen Texas counties[10] —yet none of the findings justified extensive commercial development.

In 1886, a temporary decline in Pennsylvania field production revived interest in the Oil Springs area in Nacogdoches County. During the next two years, this region experienced a brief but transitory burst of activity that marked Texas' first oil-field development. Four oil exploration companies drilled a total of ninety wells, and one of these companies, the Petroleum Prospecting Company, constructed the first crude oil pipeline and the first steel storage tanks in Texas. The pipeline, three inches in diameter and fourteen and one-half miles long, connected wells at Oil Springs with the steel storage tanks at the town of Nacogdoches. Another company, the Lubricating Oil Company, constructed the state's first oil refinery near Oil Springs and manufactured small quantities of household naphtha and illuminating oil. An estimated 150 men were employed in the area at the height of this development.[11]

By the end of 1889, however, this field was virtually abandoned. Production had diminished; drilling crews had moved on to more promising locations; refinery and pipeline installations had been dismantled or allowed to rust away in idleness. The development of more prolific oil-producing areas in West Virginia, Ohio, and Indiana brought an abrupt end to further field development. Though Oil Springs was the first commercial oil field in the state, its short life contributed little to the permanency of the Texas petroleum industry. Its production from 1886 through 1889 is unknown, although Charles A. Warner estimates it to have been less than 4,000 barrels.[12]

10. Charles A. Warner, *Texas Oil and Gas Since 1543* (Houston: Gulf Publishing Company, 1939), p. 10.
11. *Ibid.*, pp. 11–16, 81; Carl C. Rister, *Oil! Titan of the Southwest* (Norman: University of Oklahoma Press, 1949), pp. 6–8.
12. Warner, *Texas Oil and Gas*, p. 366.

The United States Geological Survey first recognized Texas as an oil-producing state in 1889, the year that the Oil Springs field went into decline. The extent of that decline was evident in the Geological Survey report, which showed that only forty-eight barrels of crude oil were produced in the entire state in 1889. This meager production came entirely from a well on the ranch of George Dullnig, near San Antonio, in Bexar County. Since the discovery of oil on his property in 1886, Dullnig had carried on a modest "oil trade." He sold the thick, unrefined crude oil as a general-purpose lubricant for from twenty to thirty-five cents a gallon. Natural gas was also found on the ranch, and the ingenious Dullnig used it for heating and lighting purposes. He said that the gas had replaced such conventional fuels as wood and coal, which normally cost an estimated $1,728 yearly. This amount the report assigned as the entire annual value of the Texas natural gas industry.[13] In fact, only the Dullnig ranch oil and gas were noted in determining Texas petroleum production from 1889 through 1895. The Geological Survey report for 1895 noted exploratory activities at Corsicana and Sour Lake but concluded rather pessimistically that "as yet this state has assumed no importance as a producing territory. . . . Practically all of the oil produced in the state in 1895 was from the wells of Mr. George Dullnig, near San Antonio, in Bexar County."[14]

Thus, as Cullinan well knew, Texas petroleum development was desultory and disappointing before Corsicana initiated extensive commercial activity. Yet, as he prepared to assess that field, he probably was not fully cognizant of the zeal and determination with which a few native Texans had continued their search for oil during these pre-Corsicana years. Signs of the existence of petroleum in many areas of the state were unmistakable. Over two hundred miles to the southeast, a man named Patillo Higgins had already drilled two test wells, in 1893 and 1895, along the Gulf Coast at

13. U.S., Geological Survey, *Mineral Resources of the United States, 1889, Non-metals* (Washington: U.S. Government Printing Office, 1890), II, 307. Hereinafter cited as U.S., *Mineral Resources.*

14. U.S., *Mineral Resources, 1895,* III, 701. Dullnig production, 1889 through 1895: 1889: 48 barrels; 1890: 54 barrels; 1891: 54 barrels; 1892: 45 barrels; 1893: 50 barrels; 1894: 60 barrels; 1895: 50 barrels.

Spindletop Mound near Beaumont. Handicapped by poor equipment and inadequate financing, Higgins had been forced to abandon his explorations.[15] But these initial failures did not lessen his conviction and that of other Texas petroleum pioneers that other areas of the Lone Star State also held riches of oil.

To Cullinan, accustomed to the craggy, forested country of western Pennsylvania, the rolling, cotton-growing prairie country surrounding Corsicana, county seat of Navarro County, must have appeared a very unlikely place to find oil. The town was primarily a cotton-processing and trading center in the rich black-belt agricultural territory of North Central Texas. Its population in 1890 was 6,285.[16] Since its settlement in 1849, Corsicana had depended, for growth and prosperity, on cotton. Cotton, and its accompanying business structure, brought to Corsicana permanent settlement by planters in the antebellum years; it attracted the railroads in the 1870s and 1880s;[17] and it spurred the immigration of a professional and mercantile group to service the structure. High cotton prices thus brought prosperity and optimism concretely expressed in rising land values, increased trade, and further investment capital. Low cotton prices, conversely, meant pessimism, shrinking land values, and general economic stagnation. Corsicana, sharing a fate in common with dozens of other Texas "cotton towns," was tied to the staple-crop system.[18]

The prolonged agricultural depression of the early 1890s reached

15. Warner, *Texas Oil and Gas*, pp. 19–23.

16. U.S., Bureau of the Census, *Eleventh Census of the United States, 1890*, vol. I, *Population* (Washington: U.S. Government Printing Office, 1893), 444.

17. At the discovery of oil in 1894, two railroads passed through Corsicana: the Houston and Texas Central Railroad, with its line south to Houston and north to Dallas which afforded interstate connections to both Kansas City and St. Louis; and the St. Louis Southwestern Railway (the Cotton Belt), which linked Corsicana with St. Louis, northeast through Texarkana, and to the west to Gatesville and Hillsboro, Texas. A third line, the Trinity and Brazos Valley Railroad, which ran northwest to Fort Worth and south to the Houston-Galveston area, was not constructed until 1905. S. G. Reed, *History of Texas Railroads* (Houston: St. Clair Publishing Company, 1941), pp. 208, 403, 413, 415.

18. In 1880, Navarro ranked third among Texas cotton-growing counties: 12,958 bales produced in one year (1879) from 45,716 acres cultivated. U.S., Bureau of the Census, *Tenth Census of the United States, 1880*, vol. V, *Cotton Production in the United States*, Pt. I (Washington: U.S. Government Printing Office, 1883), 74.

its nadir in Texas in the low cotton market of 1894, when Texas cotton prices hit a low of 4¼ cents a pound as offered by the Dallas Cotton Exchange on December 7.[19] This posed an anguishing problem to the Corsicana business community: what could be done to diversify the area's economic life? Town business and professional leaders, concerned with their careers and frequently with interests in outlying farm land, sought solutions to this complex problem in energetic, but often wistful, planning.

In January 1894, Corsicana businessmen announced the organization of the Commercial Club for the purpose of attracting industry to the locality. Club president and local attorney James L. Autry, who was later to become a close associate of J. S. Cullinan, stressed the advantages his town possessed as an established cotton-processing center with a benevolent climate, fine schools, progressive citizens, and excellent railroad facilities. But he admitted that Corsicana had one major handicap in the spirited competition among Texas communities to attract outside capital for commercial expansion: the town's water supply was hopelessly inadequate for industrial purposes. Autry promised that the first work of the new club would be to solve this problem.[20]

A private corporation, the Corsicana Water Company, chartered in 1883, had attempted to supply both commercial and domestic consumers. It operated several shallow wells and constructed a large earthen reservoir near the southeastern edge of the town for retention of surface-water runoff. But these sources failed to provide a uniform volume necessary for commercial purposes and domestic users understandably complained about the unpalatable water from the earthen reservoir and shallow wells during the hot, dry summer months.[21]

In March 1894, the Commercial Club announced its plans for solving the water-supply problem. A new company was chartered, the Corsicana Water Development Company, with capital subscribed by local business leaders. James L. Autry, president of the Commercial Club, also served as president of the new company;

19. Dallas *Morning News,* December 8, 1894.
20. *Ibid.,* January 20, 1894.
21. *Ibid.,* January 23, 1894.

Charles H. Allyn, former mayor and local merchant, was named secretary; and James Garrity, president of Corsicana's First National Bank, was appointed treasurer.[22] The new company soon announced that it had negotiated a contract with the American Well and Prospecting Company, a partnership composed of H. G. Johnston, Emlin H. Akin, and Charles Rittersbacker, itinerant well-drillers. The contract called for the latter company to drill three deep artesian wells within the city limits of Corsicana. This firm had wide experience in well drilling and had recently completed deep wells at Marlin and Waco, Texas. Autry, as company president, confidently predicted that, when completed, the new wells would have a total daily flow of 750,000 gallons with a natural pressure sufficient to fill standpipes and storage tanks without pumping installations.[23]

Work began in the spring of 1894 on the first of these wells, located on Corsicana's South Twelfth Street, one block south of the Cotton Belt railroad which bisected the town from east to west. On June 9, 1894, when the well was at 1,035 feet,[24] the drillers noticed unmistakable signs of crude petroleum filling the shaft and slowly rising to the surface. The drillers immediately sank metal casing, to seal off the crude oil from the shaft, and continued drilling. But petroleum continued to seep up on the outside of the casing; it saturated the ground around the well shaft. Meanwhile, curious and careless townspeople crowded about the well, and two fires broke out: the first, when a bystander dropped a match on the saturated ground near the well shaft; the second, as a workman struck a spark from a forge while making equipment repairs. After the second fire, workmen constructed a ditch to run off the oil seepage to an earthen tank a few yards to the side of the well. An

---

22. *Ibid.*, March 7, 1894.

23. *Ibid.*, September 29, 1894.

24. An uncertainty, albeit a minor one, exists as to the depth at which petroleum was first encountered in this well. J. S. Cullinan (Corsicana) to Z. T. Fulmore (Austin), May 19, 1899, Louis Wiltz Kemp Papers, University of Texas Archives, Austin, is the authority for the above depth. The same depth is also given in U.S., *Mineral Resources*, 1895, III, 701; *ibid.*, 1896, III, 848. However, 1,027 feet is given as the depth at which petroleum was first noticed in U.S., *Mineral Resources*, 1897, III, 104; *Derrick's Hand-Book*, II, 172; and Warner, *Texas Oil and Gas*, p. 24.

estimated 150 gallons of crude oil flowed daily into this tank as the drillers worked on through the oil-bearing strata to the completion of the water well at 2,470 feet.[25]

The evidence of petroleum made little initial impression upon the officials of the water development company, who understandably regarded the oil as a hazardous nuisance to completion of the well. During the summer of 1894, as work progressed on this first well, company officials lamented these delays; they repeatedly promised that the original water-well drilling contract would be completed and asserted that there was no plan "of utilizing the oil recently discovered in the artesian well here."[26]

Meanwhile, two Corsicana businessmen not connected with the water development company, Ralph Beaton and H. G. Damon, sent a sample of crude oil from the well to a laboratory at Oil City, Pennsylvania. The specific laboratory where this test was conducted was not reported but, because of the subsequent interest of the Standard Oil Company in the field, it should be noted that Standard's major transportation affiliate, the National Transit Company, was located at Oil City. The tests showed that Corsicana crude oil had commercial value for either fuel or refining purposes.[27]

Beaton and Damon, sensing a rare opportunity, quickly organized the Corsicana Oil Development Company, including in the enterprise an itinerant Pennsylvanian who had suddenly appeared in Texas, a man named John Davidson. Little is known of his career, either before or after his Corsicana association. In contemporary accounts, Davidson is repeatedly referred to as an "experienced Pennsylvania oil man" or "drillman" and it was reported that, during June 1895, he made several trips east to report the Corsicana situation.[28] Charles A. Warner believed that Davidson was a roving oil scout for John H. Galey and James M. Guffey—hence the relatively quick appearance of these famous wildcatters at Corsicana in September 1895.[29]

25. Dallas *Morning News*, October 10, 1895.
26. *Ibid.*, June 15, 1894.
27. *Ibid.*, November 16, 1897.
28. *Ibid.*
29. Interview with Charles A. Warner, Houston, Texas, January 16, 1963.

Regardless of Davidson's role, the Corsicana Oil Development Company lacked capital to start immediate exploration. Instead, during the next year, the group initiated an energetic leasing program to reserve drilling rights on town lots adjacent to the well of the water development company.

Typical of these first leases negotiated by the Corsicana Oil Development Company was one acquired September 6, 1894, from A. Bunert and his wife, Bertha. Terms of the lease gave the company the right to "dig, bore, and mine for, and gather all oil, gasses, coal, or other minerals" found on the specified tracts (two town lots in southeastern Corsicana) for a term of twelve years. The Bunerts, as lessors, were promised a royalty of one-tenth of the gross proceeds of the sale of any minerals found on the land.[30]

Later leases secured after the oil development and lease competition intensified often specified a bonus payment made at the time of the agreement and a yearly lease rental, in addition to the usual royalty of one-eighth to one-tenth. Also, the lease period became shorter as the oil development progressed. For example, a three-year lease to a town lot acquired by the Corsicana Oil Development Company from John C. Blankenship and wife, recorded September 20, 1897, specified that the lessors were to be paid a bonus of $100, an annual rental fee of $23, and a one-eighth royalty.[31]

In September 1895, encouraged by favorable reports from John Davidson, Pennsylvania oilman John H. Galey visited the town and soon concluded an agreement with the Corsicana Oil Development Company to begin exploratory drilling upon its leases. This agreement specified that Galey, who later assigned one-half of his interest to his long-time partner, James M. Guffey, bear the entire cost of drilling and equipping five wells on lands leased by the Corsicana company. The Pennsylvanians, upon performance, were to receive an undivided one-half interest in all leases then or thereafter owned by the company.[32]

This famous team of wildcatters were, of course, already well known within the country's petroleum industry. Galey and Guffey

30. Warner, *Texas Oil and Gas*, pp. 25–26.
31. Deed Records of Navarro County, XCII, 505.
32. Dallas *Morning News*, October 2, November 16, 1897.

had pioneered in the exploration of the Southern Pennsylvania-West Virginia field in the 1870s, the Ohio-Indiana field in the 1880s, and the Neodesha, Kansas, field in 1893. Later, backed by the Mellon banking interests of Pittsburgh, they were destined to play yet another leading role in the discovery of the Spindletop, Texas, field in 1901.[33]

Guffey and Galey characteristically wasted little time in sending drilling rigs and crews to Corsicana. Their first well was drilled on a town lot 200 feet south of the water company's artesian well. It was completed October 15, 1895; and at a depth of 1,040 feet, it initially produced two and one-half barrels of crude oil daily. The second well, drilled 700 feet from the artesian well on South Eleventh Street, produced no oil—a "duster." Drilling activity was then suspended during the winter months, but in May 1896, Galey and Guffey completed a third well at Fourth and Collins streets in east Corsicana, which at first produced twenty-two barrels of crude oil daily. The contract was completed with the drilling of the fourth and fifth wells on July 1 and August 5 of the same year. Each of these wells initially produced twenty to twenty-five barrels daily.[34]

Although Galey and Guffey thus received a one-half share in the assets of the Corsicana Oil Development Company, they soon lost interest in the field and allowed company affairs to be directed by the local manager, Ralph Beaton. Perhaps these Pennsylvania wildcatters were disappointed with the results of the test wells: production, while steady and promising at shallow depths, could hardly be called "flush." As the field developed in the following months, they became impatient at the reluctance with which local backers of the Corsicana Oil and Development Company came forth with the additional and more substantial capital needed for further leasing and field development. Their oil scout, John Davidson, sold his interest in the development company for $14,000 in June 1897, and on September 10, 1897, the restless Galey and Guffey also sold their interest, accepting $25,000 in long-term notes from

33. For contemporary biographical sketches of this team see *Derrick's Hand-Book,* I, 670, 845.

34. Warner, *Texas Oil and Gas,* pp. 26–27.

three Corsicana businessmen, S. W. Johnson, Fred Fleming, and Ralph Beaton.[35] Moreover, this rapid disillusionment with the Corsicana venture was communicated to eastern circles. In 1898, Derrick's influential *Hand-Book of Petroleum*, published at Oil City, Pennsylvania, described Guffey's attitude upon leaving the field:

> The easy-going Texan's way of doing business was not congenial to him; dash and push had characterized his operations from the very first and he had not then, nor now, reached the point in life when he was content to travel by freight train if there was an express or flyer to be had.[36]

With the sale of the Galey-Guffey interest, the Corsicana Oil Development Company, a partnership, was dissolved. A new company, a corporation, was then organized and named the Southern Oil Company. Its authorized capital was $100,000, and its major stockholders were local men: Ralph Beaton, S. W. Johnson, and Fred Fleming. Beaton, general manager of the old company, was appointed president. The consequences of the Galey-Guffey departure were considered lightly as newspaper reports proudly proclaimed that the new company was now a "home enterprise" and that its policy was to "push development and the work of boring as many wells as possible as fast as the work can be done."[37]

The results of the Galey-Guffey test wells spurred other Corsicana businessmen to action. Several petroleum development companies were quickly organized and modest capital was found for leasing and drilling ventures. Among the most active of these early companies was the Texas Petroleum Company, backed by Attorney James L. Autry, Banker James A. Garrity, and Mayor James E. Whiteselle, a lumberman.[38] The Oil City Development Company, the Lone Star Petroleum Company, and the Consumers Oil Company were also prominent in the early development of the field. In addition, town-lot owners in Corsicana quickly entered into direct

35. Dallas *Morning News*, October 2, November 16, 1897; Warner, *Texas Oil and Gas*, p. 26.

36. *Derrick's Hand-Book*, I, 846.

37. Dallas *Morning News*, October 18, 1897.

38. *Ibid.*, November 13, 1897.

negotiations with drilling contractors for exploration of their properties. Moreover, as the oil activity expanded into rural areas outside Corsicana, landowners such as the Mills, Mirus, Hardy, Walton, and Jester families often preferred to develop their properties through direct agreement with drilling contractors. The decisiveness and alacrity with which Corsicana residents entered into the early leasing and production functions of field development were noted particularly by oil historian and attorney James A. Veasey, long-time general counsel of the Carter Oil Company. After examination of the early lease records at Corsicana, he commented that "Texas people plunged into the industry in far greater numbers than did local people either in Kansas or the Indian territory."[39]

The concentrated area of exploration and production was in the eastern section of Corsicana, near the successful Galey-Guffey tests of 1895–1896. Five other producing wells were completed later in 1896 and, at the close of that year, Corsicana field production was reported as 1,450 barrels of crude oil. While the amount was minuscule in comparison with the Pennsylvania field production of over 20,000,000 barrels of crude oil in 1896, it still prompted the United States Geological Survey to report that this increase from a total Texas production of 50 barrels in 1895 was "by far the greatest increase in any one state for the year."[40]

The next year was one of greatly accelerated development: fifty-seven wells were completed and forty-three were still producing at the close of the year. As a consequence, production spurted to 65,975 barrels for 1897 in a field still largely limited to several hundred acres within the eastern portion of Corsicana.[41]

In 1898, the field expanded from the town limits to include broad areas of the countryside. The producing area began a mile or so to the northeast of Corsicana and curved to the southeast for five miles in a crescent that included the entire eastern half of the

39. James A. Veasey (Ann Arbor, Michigan) to Charles A. Warner (Houston, Texas), April 17, 1944, Charles A. Warner Collection, Texas Gulf Coast Historical Association Archives, University of Houston Library, Houston, Texas.
40. U.S., *Mineral Resources*, 1896, III, 848.
41. U.S., *Mineral Resources*, 1897, III, 102.

town and approximately a square mile of the adjacent countryside. At the close of 1898, this field numbered 342 producing wells and the total production for the year was 544,620 barrels.[42] In 1899, production increased to 669,013 barrels from 425 producing wells; and in 1900, following further eastward expansion of the field, production totaled 836,039 barrels, or "a little less than one-ninetieth of the entire production in the United States."[43]

These annual increases in production were accomplished with a hit-or-miss, oil-is-where-you-find-it attitude. Knowing little, caring little, about techniques of geologic exploration and conservation, and guided by the "rule of capture," Corsicana operators drilled as close as possible to existing wells in a feverish race to exploit production.

Here was a situation that the trained eye of Joseph S. Cullinan recognized and appreciated at once, for the so-called "rule of capture" had dominated the legal climate surrounding the early development of the American oil industry. The rule was first applied in 1889 by the Pennsylvania Supreme Court in *Westmoreland Natural Gas Company* v. *Dewitt,* and it had been universally followed by other petroleum-producing states. The "rule of capture" held that oil, like wild animals under the old English common law, was a migrating and fugitive substance of no fixed place, and hence belonged to him who "captured" or gained possession of it.[44] As evidenced at Corsicana, each operator or landowner feverishly hastened to drill and produce the oil under his land before it was drained off by adjacent operators.

Thus the producing leases, particularly within the Corsicana town limits, became a thick forest of wooden drilling derricks. A

42. U.S., *Mineral Resources,* 1898, III, 106; Dallas *Morning News,* August 8, 1899.

43. George I. Adams, *Oil and Gas Fields of the Western Interior and Northern Texas Coal Measures and of the Upper Cretaceous and Tertiary of the Western Gulf Coast,* U.S. Geological Survey Bulletin No. 184 (Washington: U.S. Government Printing Office, 1901), pp. 55, 61–62; George C. Matson and Oliver B. Hopkins, *The Corsicana Oil and Gas Field, Texas,* U.S. Geological Survey Bulletin No. 661F (Washington: U.S. Government Printing Office, 1918), pp. 213, 247.

44. For further discussion of the "rule of capture," see Henrietta M. Larson and Kenneth Wiggins Porter, *History of Humble Oil and Refining Company: A Study in Industrial Growth* (New York: Harper and Brothers, 1959), pp. 16–17.

photograph of the early producing area in eastern Corsicana, probably taken about the time Cullinan first arrived to inspect the field, shows sixty-six identifiable derricks crowded into an area of three city blocks.[45] When the number of wells inevitably brought a decrease in production to that immediate area, operators then "wildcatted" or "leap-frogged" slightly ahead to new leases, in the hope of tapping fresh pools. If these tests were successful, the cycle of indiscriminate drilling and crowded well conditions was repeated.

Corsicana operators, in their zeal to exploit the area rapidly, were further aided by technological innovation in the drilling process, which utilized the fortuitous geological characteristics of the field. Oil was found in shallow sands, geologically identified as the Upper Cretaceous and Taylor Marl, at depths of 900 to 1,200 feet, underlying a topsoil of black calcareous clay and soft rock strata.[46] A well could be put down in less time and at less expense by using drilling equipment of less weight and cost than that normally used in the older oil-producing areas of the country. Thus, the early introduction at Corsicana of the so-called rotary drill, which met these economic and geologic demands, hastened the pace—and the disorder—of field development.

The rotary-drilling method had been widely used during the 1870s and 1880s by water-well contractors throughout the United States. Its equipment was characterized in its simplest form by a rotary or revolving platform which held and rotated into the earth a length of pipe with a cutting edge or bit attached. Power was initially supplied by draft animals harnessed to move in a circle, turning the rotary platform. In the drilling of shallow-depth wells through a soft or porous subsurface, this equipment soon proved to be a rapid and cheap method of sinking a well shaft. By the late 1880s, many water-well contractors in the Dakotas, Iowa, Kansas, Texas, and Louisiana used rotary rigs in preference to the older cable-tool drilling equipment which employed gravity and percussion to hammer a well shaft into the earth.[47]

45. Adams, *Oil and Gas Fields of the Western Gulf Coast*, plate VI, opposite p. 54.
46. Warner, *Texas Oil and Gas*, p. 150.
47. For a detailed study of the development of the rotary and cable-tool systems, see J. E. Brantly, "Hydraulic Rotary-Drilling System," *History of Petroleum En-*

Two brothers, M. C. and C. E. Baker, itinerant water-well drillers lured to Corsicana by the oil-field excitement in 1895, brought the first rotary-drilling equipment to that area. They had over sixteen years' experience in operating rotary devices in Louisiana, Texas, Minnesota, Iowa, and the Dakotas. Moreover, throughout these years, the Bakers had constantly tinkered with their equipment, refining and innovating. In 1882, drilling for water in the Dakota Territory, they originated the idea of using hydraulic principles to speed drilling. A stream of water was funneled down the outside of the drill pipe; perforations in the drilling bit then allowed loose dirt and rock to be washed through the hollow pipe to the surface. Later, the brothers found that water funneled down the drill pipe and through the bit was even more effective in washing loosened material to the surface. The problem of maintaining sufficient pressure to flush water through the drilling pipe was solved by using pumps. In their early operations, before their use of steam or internal combustion engines, the Bakers erected windmills to pump their crude but effective hydraulic systems.[48]

Soon after the Baker brothers' rotary equipment had demonstrated its suitability to Corsicana conditions, other drilling contractors jettisoned their cable-tool equipment and accepted the newer drilling technique. By the summer of 1898, all the cable-tool drillers had converted to the rotary or moved on to other fields.[49]

The Bakers, meanwhile, wasted little time coming to an agreement with a local concern, the American Well and Prospecting Company. They authorized the firm to manufacture and sell rotary-drilling equipment. American Well and Prospecting was the company that drilled the water well in June 1894, leading to the discovery of oil at Corsicana. The partners of the firm, Messrs. Johnston, Akin, and Rittersbacker, remained in Corsicana and, at the time of the Bakers' arrival, they operated a small machine shop servicing drillers of the area. After successfully defending itself against

gineering, edited by D. V. Carter (Dallas: American Petroleum Institute, 1961), pp. 272–452; J. E. Brantly, "Percussion–Drilling System," ibid., pp. 133–265.

48. Brantly, "Hydraulic Rotary-Drilling System," pp. 295–296.

49. Dallas Morning News, August 25, 1898.

several patent-infringement suits, the American Well and Prospecting Company subsequently prospered and expanded its Corsicana plant as it became a major supplier to the Southwestern petroleum industry.[50] In 1944, the firm became a subsidiary of the Bethlehem Steel Corporation and operated until 1960, when the subsidiary was liquidated and its Corsicana shops closed.[51]

Thus the ease with which rotary-drilling equipment penetrated the soft subsurface to shallow oil-producing strata meant relatively modest drilling and well-completion costs. A drilling speed of 1,000 feet in thirty-six hours was not uncommon and total drilling costs averaged only $550 to $650 per well. In comparison, drilling costs in Pennsylvania fields in 1898 and 1899 averaged $1,790 to $2,140 per well. After Joseph S. Cullinan had been in Texas about a year, he estimated that the average cost of wells fitted ready for operation at Corsicana was $1,500 per well and this amount included pumping equipment, well-side storage facilities, and collecting connections.[52] Corsicana businessmen, with their limited access to speculative capital, could venture, nevertheless, comparatively modest amounts in this first phase of field exploration and development.

But these local operators soon found their visions of riches clouded by the realities of the oil business. There was no question that Corsicana crude oil had commercial potential. Tests showed that refined Corsicana crude yielded 54.5 percent illuminating oil, or kerosene, and 10.8 percent household naphtha. A government report concluded that for commercial purposes the Texas crude oil possessed a "very superior quality . . . and compares very favorably with Pennsylvania oil. . . . "[53] Yet, to realize this great economic potential, Corsicana businessmen had to confront the complex problems of storage, transportation, marketing, and manufacturing. A few local operators bravely predicted that these problems

50. Brantly, "Hydraulic Rotary-Drilling System," pp. 300–301.

51. Ava Taylor, *Navarro County: History and Photographs* (Corsicana: Taylor Publishing Company, 1962), p. 177.

52. Matson and Hopkins, *Corsicana Oil and Gas Field*, p. 250; *Derrick's Hand-Book*, II, 177; J. S. Cullinan (Corsicana) to Z. T. Fulmore (Austin), May 19, 1899, Kemp Papers.

53. U.S., *Mineral Resources*, 1899, III, 151; *Ibid.*, 1896, III, 848–849.

would be solved quickly and simply by the establishment of several refineries at Corsicana. As early as October 1895, when Galey and Guffey were about to start their first exploratory well, a newspaper ironically predicted this naive conception by reporting "if the supply [of oil] is found to be as great as is anticipated, refineries, etc., soon will be erected . . . [and] Corsicana will become a great oil center."[54] Two years later, local operators were a bit sadder and wiser; they recognized local limitations in raising the extensive capital and recruiting the trained personnel that such projects required. Yet under the flush of "oil fever," they could not restrain their enthusiasm for directing their limited capital and energies almost exclusively into the exploration and production functions.

A few experienced oilmen visiting the Corsicana field warned local operators against rapid but unrealistic expansion. "There are too many wells sunk already in the small territory explored," Samuel A. Lewis, "an old operator of Pittsburgh," complained.[55] He pointed out the immediate need to determine the extent and potential of other proven areas within the field and the need to construct storage, collecting, and refining installations for existing production. Lewis's views were shared substantially by visiting oilmen H. D. Haven of New York City, William Leeky of Oil City, Pennsylvania, and W. N. Milliken of Bowling Green, Ohio.[56]

Corsicana operators were not only reluctant to raise capital for further field development from local sources, but seemed equally hesitant to co-operate with outsiders in joint financial ventures. During the first two years of the field's existence, five outside groups evidenced interest in erecting a refinery at Corsicana. In each case, it was proposed that local businessmen supply one-half the required capital, and, in addition, the outsiders often asked sizable promotional fees or bonuses in cash or stock for their part in shaping the enterprise. These payments particularly repulsed local operators, and nothing came of the plans. A syndicate headed by Samuel M. ("Golden Rule") Jones, oil equipment manufacturer and mayor of Toledo, Ohio, offered to construct a $25,000 refinery of 1,800-

54. Dallas *Morning News*, October 10, 1895.
55. *Ibid.*, November 6, 1897.
56. *Ibid.*, November 10, December 21, December 24, 1897.

barrel capacity and advanced a plan which proposed that local operators contribute half of the capital, not in cash, but in crude oil valued at the existing market price. This proposal was also rejected. A critical newspaper pointed out that Corsicana oilmen seemed "not disposed to offer any inducement other than the oil field itself offers," and it concluded that if refineries eventually came to Corsicana it would be "without monetary inducement from the people here."[57]

Meanwhile, some Corsicana operators abetted this independent attitude by achieving early successes in locating fuel oil markets for the sale of unrefined petroleum. Several of the larger producing companies, such as Southern Oil, Texas Petroleum, Oil City, and Consumers Petroleum, constructed a few small storage tanks and gathering lines, hired sales agents, leased railroad tank cars, and entered into a brisk search for new fuel markets in Texas. A few modest sales only heightened their optimism that there were unlimited profits in this market. In late 1897, an official of the Southern Oil Company confidently predicted that in the ensuing year of operations it would earn through sales of unrefined crude oil a net profit of over $45,000 on total invested capital of the same amount.[58] Such thoughts of an annual 100 percent return on capital must have indeed made Southern Oil investors apathetic toward those who talked of substantial local contributions for refinery projects.

But Corsicana operators soon found these hopes of substantial profits from fuel oil sales dashed by the inelastic quality of that market—the number of consumers using that product in the Southwest was then extremely limited. That potentially the fuel oil market could be expanded was apparent: Texas manufacturers and railroads had long been tied to high fuel costs, particularly for coal. Plentiful Texas-produced coal and lignite, of inferior fuel qualities, then sold for $1.40 to $2.10 a ton. The price of better-grade bituminous, produced only outside the state, reflected extensive transportation charges and sold at $4.00 to $6.00 a ton.[59] But while

57. *Ibid.*, January 23, March 25, 26, 1898.

58. *Ibid.*, November 11, 1897.

59. Joseph A. Taff, *The Southwestern Coal Field,* U.S., Geological Survey, *Twenty-Second Annual Report* (Washington: U.S. Government Printing Office, 1902), III,

potential fuel oil consumers in the Southwest were obviously interested in an efficient low-cost fuel, they had little or no experience with oil. They were understandably reluctant to spend sizable sums on equipment conversion until they knew more of petroleum's characteristics and until they could be assured that a constant supply would be forthcoming.

Corsicana businessmen thus quickly became fuel oil missionaries, tramping the state and extolling the virtues of their product. Corsicana Mayor James E. Whiteselle, lumberman-turned-oilman, led his group forth to combat the forces of economic ignorance. Speaking before a business group in Dallas, Mayor Whiteselle lamented that so few Texans realized the advantages of crude oil as a fuel and declared that Corsicana oilmen planned "to educate the state on its clean, cheap qualities."[60]

But to promote this education, Corsicana oilmen, increasingly mindful of their market inadequacies, were forced to sell their crude oil at prices far below those in other petroleum fields of the country. The prevailing selling price of Corsicana crude oil was fifty cents per barrel, shipping costs paid by the purchaser. Meanwhile, Pennsylvania field crude oil, with which the Corsicana oil compared in chemical qualities, sold at 91 cents to $1.29 a barrel.[61]

Yet this low price of Corsicana crude, with its promise of substantial fuel cost reductions, was unquestionably the major factor which stimulated a few pathfinding Texas businesses to convert to oil. A certain J. Ripey, brick manufacturer, of Denton, Texas (located 120 miles northwest of Corsicana), said that even with thirty-cents-per-barrel transportation charges and the cost of converting his plant from coal to oil-burning equipment, he was still confident of at least a 20-percent saving in annual fuel expenditures.[62]

Corsicana operators were soon able to persuade other industrial consumers of the advantages of crude oil as fuel. The Southern Oil Company reported contracts with the San Antonio Gas and Elec-

291, 410; U.S., Bureau of the Census, *Eleventh Census of the United States: 1890, Manufacturing*, II, 489.

60. Dallas *Morning News*, December 29, 1897.

61. *Derrick's Hand-Book*, II, 102.

62. Dallas *Morning News*, January 8, 1898.

tric Company, the Standard Electric Company of Dallas, the Waco
Ice Company, the Dallas Gas Company, and one unnamed Dallas
manufacturing establishment that "concluded to substitute Texas
oil for foreign coal." The Texas Petroleum Company claimed shipments of two to three tank cars daily during a ten-day period in
December 1897; and the Oil City Company reported fuel oil orders
from Calvert, Texas, "on hand for several days ahead."[63]

Yet these modest ventures into the oil market did little to bring
stable field development. News of these initial sales only spurred
increased activity by smaller operators following the familiar pattern of indiscriminate drilling and crowded well-spacing conditions. If production were obtained, these smaller operators hoped
to sell the crude oil to a larger operator with a fuel oil contract. Yet
the larger operator usually had sufficient crude oil in storage from
his own wells to take care of these commitments. In late 1897, there
were thirty-four producing wells in the field contributing an additional 600 to 700 barrels of crude oil each day to storage accumulations. The two companies most active in fuel oil sales, Southern
Oil and Texas Petroleum, which claimed the largest storage facilities in the field (5,000 and 4,000 barrels, respectively), reported
their tankage filled. Total storage accumulations for the entire field
at the close of the year would reach a new high of 12,000 barrels.[64]

With storage facilities filled to capacity, field development was
soon threatened by intentional petroleum wastage. Many smaller
operators, lacking a market outlet, faced with full storage tanks,
perhaps resentful of the fuel oil contract negotiated by a neighboring producer, nevertheless allowed their wells to flow. Reports from
Corsicana told of crude oil saturating the earth around overflowing
tanks and of wooden tanks, weakened by holding a long-standing
capacity, suddenly giving way in a cascade of petroleum.[65] The
Dallas *Morning News* could contain itself no longer. In November
1897, it summarized the Corsicana field conditions under the woeful
headline: "A Terrible Waste of Oil!" as it pointed out the obvious

63. *Ibid.*, December 19, 1897; January 8, 12, 14, 1898.
64. Matson and Hopkins, *The Corsicana Oil and Gas Field*, pp. 244–247; Dallas
Morning News, January 5, 7, 1898.
65. Dallas *Morning News*, October 25, 1897.

moral that such conditions resulted in "thousands of gallons [of crude oil] running to waste."[66]

As these market and storage problems closed in on them, frustrated Corsicana operators periodically accused the railroads of failing to supply sufficient tank cars for the fuel oil trade. Both railroads serving the area, the Houston and Texas Central and the St. Louis Southwestern Railway (the Cotton Belt), admitted tank car shortages but replied that the producers often contributed to scheduling problems by giving little advance notice of impending car needs. Nevertheless, the Cotton Belt soon announced construction at its Pine Bluff, Arkansas, shops of three 8,000-gallon-capacity cars especially for Corsicana fuel oil shipments, and the Houston and Texas Central allocated seven new tank cars for this market. Yet oilmen complained that these cars were "but a drop in the bucket" and that several lucrative fuel oil contracts were lost because of shipping delays.[67]

This prevailing mood of frustration and resentment which overshadowed the Corsicana field was deepened further by the gnawing realization that another economic opportunity was escaping local operators through their failure to utilize the substantial pockets of natural gas encountered in drilling for oil. Recognizing that natural gas usually collects in reservoir formations above oil deposits, Corsicana drillers considered such gas pockets a nuisance to well completion and allowed the gas to "flare" or waste away before quickly pushing on into oil-bearing strata. Local operators, like most other Texas oilmen of the next two decades, knew little, or cared little, about the major role natural gas plays in maintaining favorable reservoir pressure for maximum oil production. By their indiscriminate waste of natural gas, Corsicana oilmen lost not only the immediate economic gain from an irreplaceable product; they also lost reservoir pressure furnished by the impounded gas, and thus lessened the petroleum production potential of their field.[68]

66. *Ibid.*, November 28, 1897.
67. *Ibid.*, November 28, December 8, 1897.
68. For a further discussion of natural gas and the problem of reservoir conservation, see Warner, *Texas Oil and Gas*, pp. 132–134.

Late in 1897, however, several Corsicana residents ventured to use the natural gas as a household fuel and illuminant. George W. Hardy piped gas from a well on his own land to his residence and was reported as "enthusiastic" over the results of the trial. Two larger producers, the Oil City and Corsicana Oil Development companies, talked of a joint venture to supply other local consumers.[69] But these plans were soon abandoned. With the growing uncertainty of field marketing conditions, these firms could not consider seriously the sizable subsidiary expenditures for natural gas collection and distribution installations.

As he concluded his inspection of the Corsicana oil field in the fall of 1897, Joseph S. Cullinan, recalling his past experiences in eastern fields, realized that this field had reached the end of its first phase of development. However, as he also recognized, there were important differences in the early development of this isolated Texas field located over a thousand miles from older petroleum centers. Despite limitations of capital and petroleum experience, local businessmen, aided by a union of favorable geological circumstance and technological innovation, had dominated early production ventures. Yet, as Cullinan had seen in his Pennsylvania days, those initiating this function were often beguiled by the feverish pace of exploration and failed to face the realities demanded by the next phase of field development. Solution of such problems as the need for storage facilities, gathering lines, refinery construction, and additional market outlets demanded extensive marshaling of capital and managerial talent. Cullinan well understood the plight of the Corsicana oilmen, for their limited capital had been expended by the first phase of development and their visions of oil riches were now clouded by their lack of experience in the complexities of the petroleum industry. Desperately seeking this all-inclusive help, local business leaders had turned to "outsiders." In their invitation to aggressive Joe Cullinan, who was polished within the managerial structure of the Standard Oil Company, Corsicana businessmen were soon to find solutions to their problems.

69. Dallas *Morning News*, October 6, 1897, January 7, 1898.

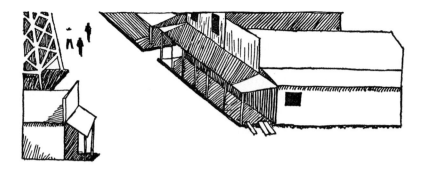

2. The Corsicana Oil Field: Cullinan Begins Consolidation, 1897-1899

As JOSEPH S. CULLINAN surveyed the Corsicana oil field in the fall of 1897, it was obvious that the first step toward solution of the field's mounting problems would involve construction of adequate storage facilities. Furthermore, as the managing partner of a firm specializing in tankage construction, Cullinan obviously could meet this need quickly, regardless of his future decisions involving the area's petroleum development. In early December 1897, after a short visit to Washington, Pennsylvania, Cullinan returned to Corsicana with a crew from his Petroleum Iron Works. Heavy steel plate and other equipment followed, and construction soon began on the field's largest storage facility, a 16,000-barrel tank eighty feet in diameter and twenty-five feet in height, erected beside the Cotton Belt railroad tracks in southeastern Corsicana.[1]

Cullinan's initial inspection of the field had been closely watched

1. Dallas *Morning News,* December 2, 4, 11, 1897.

29

by the local citizens, but he had been inscrutable as to his future plans. Now that the tank construction evidenced his concrete interest in the field, local excitement and rumor accelerated. Scores of questions were put to Cullinan: Did this mean a refinery was planned for Corsicana? If so, where was it to be located? Who was backing Cullinan in his venture?—But Cullinan, as newspaper accounts noted, was characteristically "close-mouthed" and did not "let go what they wanted to know."[2]

The Pennsylvanian chose to remain silent for several weeks as to the extent of his plans. But on the eve of a Christmas holiday trip to visit his family in Pennsylvania, Cullinan did admit that, while in the East, he planned to complete arrangements for additional tankage and a complete pipeline gathering system to link producing wells with centralized storage facilities. However, when pressed as to his intentions about refinery construction, Cullinan was rather vague. He promised only to consult with "his principals" as to advisability of a "refinery proposition." Nevertheless, oil circles were reported as jubilant over this news that promised an immediate solution to the field's storage problem and to the implication "that in a comparatively short time there will be right here in Corsicana a cash market for the total output of the oil wells producing and for the output of any additional wells."[3]

Upon his return to Corsicana in January 1898, Cullinan soon revealed his further plans for the oil field. He announced a contract with the three largest producers in the Corsicana field: the Southern Oil, Texas Petroleum, and Oil City companies. Cullinan was to purchase 100,000 barrels of crude oil from these companies at a price of fifty cents per barrel. It was further specified that Cullinan need not take or buy more than 1,000 barrels of crude oil a day, unless, at his option, he elected to receive and pay for additional amounts. The contract was to be completed within a period of two years.[4]

Although this contract gave Cullinan control of the output of the field's three major producing companies, he was careful to point out that its terms would also be extended to the smaller pro-

2. *Ibid.,* December 3, 1897.
3. *Ibid.,* December 22, 1897.
4. Deed Record of Navarro County, XCII, 629–30.

ducers in the field. Since the three major companies were then producing 700 to 800 barrels daily, additional purchases would be made from smaller operators up to the 1,000-barrel daily limit, or beyond, as storage and collecting facilities were expanded. Cullinan also said that, while the price of fifty cents a barrel was to stand for the duration of the 100,000-barrel contract, later marketing and production conditions might justify higher prices for local producers. But in response to further questioning, the Pennsylvanian admitted that he had no plans at present to construct a refinery. His purpose in Corsicana, he affirmed, was to construct a field storage and collection system and to offer the purchased crude oil "at a uniform price of 70-cents per barrel to whoever wishes it."[5]

But not all Corsicana oilmen were enthusiastic over this news of Cullinan's entrance into the local field. Suspicion and doubt immediately arose as to the eventual fate of the smaller producers under the 100,000-barrel contract. The Dallas *Morning News* correspondent in Corsicana predicted that Cullinan's contract with the major producers and his control of field storage facilities would force smaller producers to sell their oil to him. He reported that, while Cullinan's plans for the local field had excited local interest, "the people hardly know whether it is a good thing or not." But the newspaper later seemed willing to reserve immediate judgment on the problem by concluding, somewhat cryptically, that the fate of the "small fry on the outside [of the 100,000-barrel contract] is a question left for them and other people to answer along with them. . . ."[6]

Further suspicion soon arose as rumors swept Corsicana concerning the nature of Cullinan's financial backing. It was soon charged that he was an agent of the Standard Oil Company. Fears were expressed that, because of his presence, the oil industry would slip out of the hands of local people into the hands of those who had no particular interest in the building up of Corsicana and who "would not so manipulate the oil output as to make it operate to the general good of this city."[7]

Cullinan did little to dispel these rumors during his first months

5. Dallas *Morning News*, January 18, 1898.
6. *Ibid.*, January 7, 18, 1898.
7. *Ibid.*, January 7, 1898.

at Corsicana. In a newspaper interview at the time of the announcement of the 100,000-barrel contract, he was asked directly whether
he represented the Standard Oil Company. Cullinan was quoted as
replying: "I am representing myself and capital back of me, but I
am not making this deal for the Standard Oil company. Still, this
contract will not conflict or be inimical to the interests of the
Standard people."[8]

Thus, from the beginning of his Corsicana association, it was
widely believed that Cullinan, in some way, was connected with the
Standard Oil Company. Local acceptance of this situation was well-
evidenced by a newspaper article commenting on a United States
Geological Survey report concerning the Corsicana oil field development. It was pointed out that the government's report con-
tained several errors. Among them was a statement that Cullinan
was connected with the "Southern Oil Company." The newspaper
tersely put the record straight: "He should have been credited to
the Standard Oil Company."[9]

This belief that Cullinan represented the Standard Oil involvement in the Corsicana field was indeed a matter to excite local question and concern. Under heavy attack in the national "court of
public opinion," the "Standard Trust" already faced legal pro-
ceedings in Texas to oust from the state its major marketing affiliate
in the Southwest, the Waters-Pierce Oil Company.[10] For in 1896,
Texas authorities, alleging violation of state antitrust laws, instituted action to suspend the permit of Waters-Pierce to carry on
business within the state. The prosecution was able to present a
convincing case of unfair marketing practices and won a forfeiture
judgment on June 18, 1897, in the District Court of Travis County.
This judgment was affirmed by the state's Third Judicial District
Court of Civil Appeals on March 9, 1898, and by the Texas Supreme Court on May 5, 1898. But the corporation was still doing
business in Texas at the time of Cullinan's appearance in Corsi-

8. *Ibid.*, January 18, 1898.
9. *Ibid.*, February 13, 1898.
10. For a discussion of journalistic attacks on the Standard Oil group and the early
history of state litigation against the company, see Hidy and Hidy, *Pioneering in Big
Business*, pp. 642–652.

cana, since the decision was on appeal to the United States Supreme Court.[11]

Despite these contemporary rumors linking him with Standard Oil, evidence supports Cullinan's later assertion that he initially came to Corsicana in an independent capacity. In 1913, as state authorities again filed antitrust action against Standard-controlled companies in Texas, Cullinan was subpoenaed as a witness and gave a deposition concerning his early relationship with Standard Oil at Corsicana.[12] He recounted that he arrived in Corsicana for a visit in October 1897, after having corresponded with Mayor James E. Whiteselle, and that his reason for making the visit was "the possibility of my taking an interest or interesting others in

11. For a concise statement of issues and testimony presented in the Waters-Pierce case in trial and appeal through the state courts, see Texas, *Report of the Attorney-General, For the Years 1897–98: M. M. Crane* (Austin: Von Boeckmann, Moore & Schutze, State Contractors, 1899), pp. 10, 14, 17, 19–20.

In March 1900, the United States Supreme Court affirmed the Texas judgment. In May 1900, however, Waters-Pierce was reorganized under Missouri law and was granted a permit to do business in Texas after submitting an affidavit that the company was no longer a member of an organization of the trust type. In 1906, Texas again instituted antitrust action against the company, alleging that Waters-Pierce had fraudulently concealed connections with the "Standard Trust" when seeking the re-entry permit in 1900. Texas courts rendered judgment for the state, revoked the company's permit to do business, and assessed $1,623,900 in penalties. This judgment was also affirmed by the United States Supreme Court in 1909. The company then paid the assessed penalties and sold all its property, under court order, situated in the state. For a review of this later litigation, see Texas, *Report and Opinions of the Attorney General, For the Years 1908–1910: Jewel P. Lightfoot* (Austin: Austin Printing Co., Printers, 1911), pp. 34–36.

12. Deposition dated February 15, 1913, *State of Texas* v. *Magnolia Petroleum Company et al.* (Certified copy in Cullinan Papers. The following quotations and references concerning Cullinan's role at Corsicana are from this document unless noted otherwise.)This deposition was given in a state antitrust suit against the Magnolia Petroleum Company, successor corporation to the Corsicana companies organized by Cullinan. It was alleged that the Magnolia company was secretly controlled by Standard Oil executives Henry C. Folger, Jr., and John D. Archbold, and that this control violated a 1907 Texas antitrust judgment barring Standard-affiliated companies from the state. This contention was admitted by the Magnolia company in a pretrial settlement. Standard Oil of New Jersey paid a $500,000 fine, but Magnolia was allowed to continue in business in Texas following the transfer of the Folger-Archbold stock to a court-appointed trustee. For further information on this litigation, see Texas, *Bienniel Report of Attorney General, 1912–1914: B. F. Looney* (Austin: Von Boeckmann-Jones Co., Printers, 1915), pp. 15–16.

the field's development." After making his preliminary investigations, Cullinan returned to the east and reported his findings to business associates in Washington, Pittsburgh, and Franklin, Pennsylvania. These associates were impressed with Cullinan's report and urged him to go ahead with his Corsicana plans while they would proceed to raise the necessary capital for a field storage and collecting system and, later, if conditions justified, for a refinery.

Cullinan then returned to Corsicana with these assurances and, using mainly his own funds, proceeded to construct storage and gathering facilities and to negotiate the 100,000-barrel oil purchase contract with the field's major producers. Later, when Cullinan went east again to obtain additional capital for a refinery, he found that his friends had "reconsidered and could not then see their way clear to proceed with the proposition." He added, cryptically, that this indicated an "influence had been brought to bear from some source which was not uncommon in those days."

The above statement merits special attention. Did Cullinan mean that Standard Oil exerted pressure in Pennsylvania oil circles to reserve Corsicana development for that company? This is certainly a possibility, but it would seem highly unlikely that Standard Oil, already under heavy legal and political attack in Texas and throughout the nation, would risk further criticism in overtly pressuring entry into the still relatively minor Corsicana field. In later years, Cullinan talked of his independent backing among Pennsylvania oilmen fading away because of adverse "reports in government circles" as to the future of the Corsicana field.[13] Yet, examination of such government reports as the United States Geological Survey Bulletins (circa 1895–1903) fails to indicate an adverse comment on the future of the Corsicana development or the quality of oil produced there. Further, it should be recalled that Cullinan made this statement in 1913, eleven years after the severance of his Standard Oil connections at Corsicana, and at a time when he served as president of the Texas Company, not only a major competitor of Standard Oil but a company exceedingly proud, as its name would indicate, of its development on the basis of local capital and management. One surmises, then, that in

13. Houston *Post*, April 21, 1934. Interview with J. S. Cullinan.

Cullinan's 1913 statement he could not resist thrusting a further business barb or two at Standard, while at the same time purging himself of his past associations and connections.

The major reason for the failure of his Pennsylvania friends to support his Corsicana plans for a refinery installation was that Cullinan's plans were too grandiose, that they asked too much in capital commitment. While his Pennsylvania contacts unquestionably had modest investment capital, the substantial amounts Cullinan needed for refinery installation and for the salvaging of his personal involvement in crude oil contracts was beyond their will and means. Their failure to grasp quickly the Corsicana opportunity as Cullinan saw it meant that, with time crowding in on him, he was forced to seek help from his old Standard Oil contacts.

If Cullinan's assertion that he initially sought independent help at Corsicana is to be accepted, the Pennsylvania friends whose unwillingness to support him brought Standard Oil into the field should be identified. It has been written that, at first, Cullinan had the backing of John Galey and James Guffey to construct a refinery at Corsicana. Later, the withdrawal of this team of peripatetic wildcatters caused Cullinan to seek Standard Oil help.[14] This undocumented conclusion, of course, is highly unlikely, in view of the pessimistic and irritated attitude with which this team severed their connections with the Corsicana field in September 1897.

Several years later, Cullinan confided to a close business associate that he initially had planned to obtain his major financial support from his old oil friends and contacts in western Pennsylvania. It was hoped that these independent oilmen, predominantly modest operators with backgrounds in exploration and production, would pool their resources in supporting the Corsicana venture. Cullinan particularly counted on his old friends and partners in the Petroleum Iron Works of Washington, Pennsylvania—Messrs. Todd, Wright, Slater, and Sweeney—for substantial contributions.[15] But

14. Robert C. Cotner, *James Stephen Hogg: A Biography* (Austin: University of Texas Press, 1959), p. 520.

15. Cullinan (Beaumont) to W. J. McKie (Corsicana), February 4, 1905, Cullinan Papers.

the failure of this group to raise or risk the necessary capital contemplated in the Corsicana operations obviously sent Cullinan searching for more affluent friends.

Thus, Cullinan's assertion that a legitimate effort was made initially to obtain independent or non-Standard-Oil support for his Corsicana enterprises must stand largely unchallenged, even though the story is admittedly incomplete and somewhat clouded. Furthermore, it appears that additional details will never be learned, since most of Cullinan's early Corsicana business papers for the years 1897–1901 apparently have been destroyed. In 1913, long after Cullinan had severed his connections with Standard, but at a time when Standard Oil companies in Texas again faced antitrust litigation, a company official testified that all of the early Corsicana records, including Cullinan's papers, "had been destroyed, as the room which they occupied was needed."[16]

Cullinan further stated in his 1913 deposition that, when the initial plans concerning his Pennsylvania associates failed, he immediately explained the situation fully to the Corsicana businessmen involved in the local field's development. It was agreed by all that Cullinan's personal resources, plus what little help might come from local sources, would be wholly inadequate in meeting prospective capital requirements. The situation was indeed unpromising and Cullinan, with the approval of the Corsicana group, suggested that he write his old friend and benefactor during his years of employment with Standard Oil-affiliated companies, Calvin N. Payne of Oil City, Pennsylvania. Payne was the general manager of Standard's major pipeline affiliate, the National Transit Company, and had continued his climb within the Standard hierarchy since his young protege, Joe Cullinan, had left his employ to go on his own in the oil equipment business. Calvin Payne was a member of Standard's elite inner circle of trust managers, a director of five other Standard affiliates, and very shortly (1899) to be rewarded with appointment to the board of directors of the major policy and planning unit, Standard Oil of New Jersey.[17] Irish pride

16. *Oil and Gas Journal* (June 12, 1913) p. 8, reported the testimony of W. C. Proctor, an officer of the Magnolia Petroleum Company, a successor corporation to a Cullinan-organized enterprise at Corsicana.

17. Hidy and Hidy, *Pioneering in Big Business*, pp. 229, 231, 329, 385, 648.

notwithstanding, Cullinan, fearful that his Corsicana hopes were in danger of complete collapse, felt it time to talk with his old mentor.

Cullinan met Payne in St. Louis in April 1898. A preliminary agreement was quickly reached. Payne, with others not then named, agreed to form a partnership, details to be worked out later, fully backing Cullinan's Corsicana plans.

By early May 1898, the preliminary plans had been completed and the "Articles of Co-Partnership" signed. In addition to Cullinan and Payne, the enterprise now included another partner, Henry C. Folger, Jr., of New York.[18] Folger was at that time an official of Standard Oil of New Jersey and chairman of that firm's select Manufacturing Committee. In 1908, he was to become a director of "Jersey Standard" and, following dissolution of the combine under federal court order in 1911, he became president of Standard Oil of New York.[19]

The agreement further specified that a copartnership, to be known as "J. S. Cullinan & Company," was to be formed at Corsicana. It was to conduct a general petroleum business including exploration, production, transportation, marketing, and manufacturing activities. Payne and Folger were to direct financial management of the business, but Cullinan was to have the entire responsibility for local management of the enterprise.

The initial capital of the partnership was $100,000, but the amount could be enlarged from time to time as the need of the business demanded. Payne and Folger contributed all of the capital, one-half from each.[20] Cullinan was to contribute his personal services and was to be paid an annual salary of $5,000. Payne and

18. Articles of Co-Partnership, dated May 8, 1898, between J. S. Cullinan, C. N. Payne, and H. C. Folger, Jr. (copy), Cullinan Papers. The following references and quotations pertain to this instrument unless otherwise noted.

19. Hidy and Hidy, *Pioneering in Big Business,* pp. 59, 283, 314, 319, 410, 712.

20. The capital contributions of Payne and Folger were advanced to the enterprise through the "books" of the National Transit Company. (Hidy and Hidy, *Pioneering in Big Business,* p. 393.) This arrangement was noted by Texas authorities in state antitrust actions against Standard Oil affiliates in 1907 and 1913. It served to heighten the suspicion that the Texas affiliates were closely controlled by Standard Oil's trust managers and to defeat the claims of Standard officials that the Texas companies were the result of individual initiative on the part of Folger and Payne. (*Oil and Gas Journal,* June 12, 1913, p. 8.)

Folger were to share equally all profits and losses of the partnership. With this assurance of financial support from Payne and Folger, Cullinan quickly returned to Texas and his plans for further field development. By placing local direction in the hands of the young Pennsylvanian, the astute Payne and Folger had assured that whatever turn Corsicana took, their financial backing would receive vigorous managerial support.

One criticism of the copartnership agreement, however, is that Payne and Folger might have taken advantage of Cullinan's loss of promised capital from other sources and obtained his services at Corsicana at compensation substantially below that paid comparable Standard managers. Standard Oil at that time paid modest salaries to managerial-level employees: in 1886, only forty in the entire organization of over 8,000 employees received annual salaries of $10,000 or more. By 1901, this number had almost doubled, but salaries of some staff auditors had reached $5,000 annually.[21] Cullinan's past record and the scope of his Corsicana responsibilities would seem to have justified more than the salary paid those with more prosaic duties.

But if Cullinan felt in this early period that an inequity existed, the grievance was well hidden by the subsequent managerial energy and determination he displayed on behalf of the new enterprise. Although Payne and Folger eventually raised Cullinan's pay to $8,000 in 1901,[22] at this point in his career, he was probably little concerned with the status of salary. With his family's needs comfortably provided for by the partnership salary and the additional income from his Pennsylvania company, the Petroleum Iron Works, Cullinan was more interested in the future opportunities that might arise at Corsicana once he was established there by adequate financial backing.

For Payne and Folger, the agreement of May 1898 also had immediate attraction and future promise. They were uncertain as to the climate of public and legal opinion in Texas, with prosecution of the Waters-Pierce Company already under way. Yet, since they were vitally interested in the Corsicana field's potential,

21. Hidy and Hidy, *Pioneering in Big Business*, pp. 581–582.
22. *Oil Investors' Journal*, October 5, 1907, p. 20.

these Standard Oil executives could now make an entry into that area via the trial balloon of the Cullinan partnership. While the capital required was substantial, it was by no means an overwhelming commitment, in view of the contacts of Payne and Folger. And above all, these partners were assured that, with local direction of their venture in the hands of Cullinan, the enterprise would be managed at a modest salary by an experienced and trusted associate.

The announcement in early 1898 of Cullinan's 100,000-barrel crude oil contract with the major producing companies launched another round of frenzied leasing and drilling activity. This activity was not only concentrated in the older, proven area within the town limits, but greatly expanded the field to the northeast and southwest of Corsicana. Within a month, successful wells were drilled near the small village of Chatfield, ten miles to the northeast, and along the Houston and Texas Central Railroad two miles to the southwest. At the end of January 1898, there were 65 producing wells in the field, with 17 in process of being drilled. By the end of February, the number of producing wells in the field had spurted to 80, with 48 wells in the process of being drilled. At the end of March, the number of producing wells had increased to 112, with 30 then being drilled.[23]

Moreover, operators of older wells in the field began to employ mechanical means to boost faltering production. With compressed air used to expand reservoir pressure, production of some wells within the crowded town limits of Corsicana was increased from ten to thirty barrels of crude oil daily. This method was brought to Corsicana by drillers familiar with the equipment commonly used in the older oil regions of the country. The first patent was granted in 1864 for a compressed-air device perfected in the Oil Creek, Pennsylvania, field. The equipment, in its simplest form, consisted of a hollow iron tube lowered down the well casing and then connected at the surface to a hand-pump compressor. Air was then forced down the tube and "lifted" the oil up the casing to the surface. Despite the irony that, through earlier haste and ignorance, Corsicana operators had lost much of the reservoir pressure furnished by pockets of natural gas, these operators now boasted of

23. Dallas *Morning News*, January 26, February 2, 26, March 19, April 2, 1898.

increasing the output of their wells by as much as "300%" through the use of "air compressor machines."[24]

Even before the partnership agreement of May 1898 had been completed, Cullinan had begun the construction of gathering and storage facilities necessary to service his 100,000-barrel crude oil contract. By late January, the J. S. Cullinan Pipe Line Company had connected the wells of the major producers involved in that contract—the Southern Oil, Texas Petroleum, and Oil City companies—with a 16,000-barrel tank in southeastern Corsicana. As field production increased, additional storage became necessary. A second tank of 26,000-barrel capacity was constructed in March; a third tank, to hold 36,000 barrels, was completed in April; and a fourth tank, also holding 36,000 barrels, was completed in May.[25]

Throughout these months, Cullinan bought the daily production from the companies involved in the 100,000-barrel contract and also extended gathering lines to the wells of the smaller producers not included in this agreement. Purchase of daily crude oil production was made at the same price guaranteed the larger producers, a situation which unquestionably stimulated the pace of further field expansion. Furthermore, in May 1898, Corsicana producers heard the news that, in several eastern oil fields, the price of crude oil had advanced nine cents per barrel. This news, they reasoned, immunized local producers against any price cut and indicated a prospective price increase for Corsicana crude oil in the near future.[26]

But in a few days Corsicana producers were shocked by the sudden announcement that Cullinan would buy crude oil only from the companies involved in the original 100,000-barrel agreement. Cullinan gave a lack of storage facilities as the reason for the action and promised that purchases would be resumed as additional storage was constructed. One newspaper was skeptical, noting that the cancellation of purchases from all but the major companies still

24. C. J. Coberly, "Production Equipment," in *History of Petroleum Engineering*, edited by D. V. Carter (Dallas: American Petroleum Institute, 1961), pp. 678–80; Dallas *Morning News*, February 5, 23, 1898.

25. Dallas *Morning News*, January 24, March 18, May 7, 1898.

26. *Ibid.*, May 10, 1898.

would not result in an over-all decrease in field production. The storage problem, if there was one, would remain and Cullinan's excuse, it was concluded, "falls to the ground."[27]

The next day, the smaller producers met to protest Cullinan's action. Chairman for the meeting was James A. Garrity, a banker and an investor in the Texas Petroleum Company, one of the larger companies involved in the 100,000-barrel contract. Garrity reminded the smaller producers that, under the terms of that contract, Cullinan was relieved from purchasing more than 1,000 barrels of crude oil a day. Garrity pointed out that, with recent field production ranging from 1,600 to 1,800 barrels a day, Cullinan had, for some time and at his own election, actually bought more oil than was required under the contract. Garrity concluded that Cullinan was doing the best he could under the circumstances. The meeting continued in an atmosphere of "calmness and moderation" and concluded with the appointment of a committee of the smaller producers to investigate the building of a competing pipe line and storage system[28]—a project which obviously was to be soon forgotten: for, within a few weeks, Cullinan, his capital now supplied by the Payne-Folger partnership, began the construction of additional tankage. He then redeemed his promise and resumed purchases from the smaller producers at a time when total field production was 1,800 barrels a day.[29]

But early in Cullinan's career at Corsicana, local producers had been exposed to market problems in a field dominated by one buying, gathering, and storage enterprise. The low price of fifty cents a barrel, at first reasonably attractive to Corsicana producers, obviously served to draw Cullinan to that area. As his plans expanded into a projected refining installation, it was increasingly essential to Cullinan and his partners that this low price be maintained. In January 1899, with the refinery in operation, Corsicana producers again pressed for increased crude oil prices. "I am sick and tired," stated a local operator, "of selling oil at 50 cents per barrel to the Standard Oil Company, when the same corporation

27. *Ibid.*, May 16, 1898.
28. *Ibid.*, May 17, 18, 1898.
29. *Ibid.*, June 8, 1898.

pays double that price for eastern oil that grades no higher than our own." Since Pennsylvania crude of a comparable grade was reported at that time as selling for $1.16 a barrel, and lower-grade Lima-Indiana field oil at seventy-five cents a barrel, Corsicana producers were understandably growing restive.[30] But Cullinan took his time. Two months later, following further price increases for eastern crude, the price at Corsicana was raised from fifty to sixty cents a barrel. Four months later, the price was seventy-five cents. At the end of that year, it was $1.03 a barrel. Meanwhile, Pennsylvania crude had ranged from $1.27 to $1.51 a barrel over the same period.[31]

Despite these later resentments over pricing policy, Cullinan's first entrance into the Corsicana field as a crude oil buyer met with wide approval from local producers. A guaranteed price, low as it might be, was established, and field storage problems were largely solved. Even more reassuring to local producers, remembering well their past frustrations in attempting to find a crude oil market, was Cullinan's assuming of the single responsibility for discovering further outlets for the field's production.

With financial backing for a refinery still uncertain during the first months at Corsicana, the Pennsylvanian attempted to expand the fledgling fuel oil market so uncertainly launched before his arrival. The selling price of fuel oil remained at seventy cents a barrel (fifty cents "well" price, plus twenty cents pipeage fee, transportation costs paid by buyer) and Cullinan stated that his handling of the major share of the field's production would not interfere with the completion of existing fuel oil contracts. He was shortly able to announce the successful negotiation of a new contract for the purchase of a substantial quantity of fuel oil. The Waco Ice Company agreed to purchase 2,500 barrels of Corsicana crude oil and "to substitute it for coal on account of better results obtained at a cheaper price. . . ."[32]

In a further attempt to increase the crude oil market, Cullinan persuaded the Cotton Belt railroad to lend him a locomotive for

30. *Ibid.,* January 17, 1899.
31. *Ibid.,* August 2, October 6, 1899; *Derrick's Hand-Book,* II, 105.
32. Dallas *Morning News,* January 18, February 15, 1898.

use in a fuel oil experiment. Because he knew that eastern railroads had found fuel oil very successful in recent trials and that California lines had used such fuel exclusively since 1894, Cullinan saw no reason why Corsicana crude could not be similarly utilized.[33]

Busy with a myriad of tasks out in the field, Cullinan delegated direction of the installation of oil burners in "Engine 43" to a younger brother, Dr. Michael Patrick Cullinan, who had recently arrived at Corsicana. Born in the western Pennsylvania oil region in 1865, the younger Cullinan had fallen from a hayloft in his childhood and received a crippling leg injury necessitating the use of a crutch throughout the rest of his life. He graduated from Starling Medical College, Columbus, Ohio (the genesis of the Ohio State University College of Medicine), in 1887, was licensed as a physician in his native state and for several years practiced medicine at Petrolia, Pennsylvania. But in an era when the practice of a "horse-and-buggy" country doctor demanded a high degree of physical endurance, the young man's handicap finally forced him to abandon his profession. He eventually became an employee of his older brother's Petroleum Iron Works in Washington, Pennsylvania and shortly followed him to Texas. "The Doctor," as he was understandably called in Cullinan family circles, was eventually placed in charge of the fuel oil sales of the Corsicana partnership and became Texas sales agent for the Petroleum Iron Works. Later, he would follow his brother Joe to the oil development in the Texas Gulf Coast area. Although he apparently resumed the practice of medicine in Beaumont for a few years, the oil industry had become his principal interest and, in 1910, he moved to Laredo, Texas, to serve as president of the Border Gas Company, a subsidiary of the Texas Company. He continued as head of this company until his death in 1927. In his later years, as a hobby, he studied the South Texas citrus-fruit industry and became a well-recognized expert and consultant in this field.[34]

A few days after being put in charge of the conversion of the

33. *Ibid.*, April 11, 1898; James A. Clark, *The Chronological History of the Petroleum and Natural Gas Industries* (Houston: Clark Book Co., 1963), p. 71.

34. J. S. Cullinan (Beaumont) to W. J. McKie (Corsicana), April 23, 1904, Cullinan Papers; *The Texaco Star*, XIV (February 1927), 3.

Cotton Belt locomotive, "the Doctor" had Engine 43 ready to roll, with oil burners in place under the steam boiler. On April 10, 1898, the locomotive pulled the regularly scheduled passenger train the forty miles between Corsicana and Hillsboro, Texas, "on time and without a mishap of any kind."[35] While Joe Cullinan was unable to persuade the Cotton Belt to adopt oil fuel at that time, this successful trial was well remembered. In just a few years, the discovery of the prolific Spindletop field near the Gulf Coast brought much lower crude oil prices and Texas railroads quickly converted to fuel oil. The Houston and Texas Central Railroad, which also served the Corsicana oil field area, led the way in converting to oil early in 1901.[36]

Corsicana's long, dry summers with clouds of dust swirling up from its dirt streets gave Joseph Cullinan further ideas for crude oil sales. He persuaded the city council to allow him to sprinkle crude oil on a few streets and results in controlling the dust were so successful that a contract was awarded to treat all city thoroughfares. Soon other Texas cities heard of Corsicana's dust-free streets and purchased crude oil by the tank car for similar purposes. The city of Fort Worth purchased a carload of crude oil in the spring of 1899 at ninety-one cents a barrel and continued to make additional purchases during that summer. Other carload shipments were also made to Waco, Greenville, and Honey Grove, Texas.[37] Even after Cullinan had constructed his refinery, sales of crude oil for street-treating purposes continued to contribute a minor but seasonably brisk market outlet.

Cullinan also turned his attention to the economic possibilities offered by the distribution of natural gas, which earlier Corsicana operators had considered a nuisance to oil-well completion. There had been some limited experiments in distributing the gas for commercial purposes. The Corsicana Gas and Electrical Light Company was reported to have temporarily connected its "coal gas" (manufactured gas) distributing lines to a natural gas well for a

35. Dallas *Morning News,* April 11, 1898.

36. Warner, *Texas Oil and Gas,* p. 33; Rister, *Oil! Titan of the Southwest,* pp. 67–68.

37. Dallas *Morning News,* March 20, September 3, 1899.

few days in March 1898, and found the natural product entirely satisfactory for illuminating purposes.[38] During 1899, however, five additional gas wells of commercial volume were discovered on the outskirts of eastern Corsicana. Cullinan, together with William H. Staley, a local operator who had lease rights to several of these new wells, now obtained a franchise from the city to supply local consumers. At the end of that year, enough distributing pipe had been laid to furnish natural gas for eighty-five domestic and twenty-one business establishments at rates of $1.50 to $2.50 per month for domestic use and $5.00 per month for business use. The local gas company also announced, in December 1899, that it was now changing exclusively to natural gas distribution. Corsicana thus became the first Texas city to be supplied with natural gas. The value of gas consumed from the wells was $8,000 in 1899 and $20,000 in 1900. These wells would continue to supply the city until 1913, when the large Mexia-Groesbeeck gas field was discovered thirty miles to the south. Gas from this field was then piped to Corsicana.[39]

In a few years, the discovery of extensive gas fields within the state and the formation of larger transmission and distribution companies would dwarf the Corsicana accomplishment. The late oil historian Charles A. Warner was undoubtedly correct when he maintained that the "beginning of the gas industry *as it is known today in Texas*" (emphasis supplied) can be traced to the discovery of the very productive Petrolia gas field in Clay County in 1907 and the subsequent organization in 1909 of the state's first major transmission firm, the Lone Star Gas Company, which supplied both Fort Worth and Dallas with Petrolia natural gas.[40] Yet the Texas natural gas industry, through Cullinan's determination to use the field's full economic potential, was initiated at Corsicana.

Thus, in two years at Corsicana, Joseph S. Cullinan, backed by non-Texas capital, had exercised vigorous managerial direction

38. *Ibid.*, March 27, 1898.

39. U.S., *Mineral Resources, 1899*, II, 299, 301–302; *1900*, III, 55; Dallas *Morning News*, October 31, December 18, 28, 1899; Matson and Hopkins, *The Corsicana Oil and Gas Field*, p. 213.

40. Warner, *Texas Oil and Gas*, p. 51.

to gain control of a major portion of the field's production. He had also constructed extensive gathering and storage facilities and made successful and ingenious efforts to find new markets for Corsicana crude oil and natural gas. But while he had largely consolidated the field's production, the further tasks of integration, particularly in the fuller control of production and the initiation of the manufacturing function, still awaited and challenged him at Corsicana.

---

3. The Corsicana Oil Field: Cullinan Directs Integrated Operations, 1899-1901

SIGNIFICANT AS WERE his efforts to expand the crude oil and natural gas markets, Joseph S. Cullinan realized that the field's economic potential could not be fully developed until a refinery was constructed at Corsicana. By the summer of 1898, with field production on the increase, with his gathering and storage system largely completed, and with the Payne-Folger partnership established, promising substantial capital, Cullinan was ready to enter this further phase of Corsicana's petroleum development.

A 136-acre tract near the southwestern limits of Corsicana, easily accessible to both the Houston and Texas Central and the Cotton Belt railroads by spur connections, was chosen as the refinery site. Construction began in June 1898, and through the following months, as the railroads brought in carload after carload of equipment, the complex refinery installation slowly took shape, amazing many local observers at the "great change in what a few weeks ago was a bare prairie."[1]

This was not, however, to be the state's first activity in the refining of crude oil. In the late 1880s, the Lubricating Oil Company had operated a "refinery" in the Oil Springs field, Nacogdoches County. It was a very primitive affair, hardly worthy of the name, employing an open, heat-fired evaporating pan and a cloth strainer to rid the crude oil of its impurities. A similar contrivance of a 150-barrel capacity operated at Sour Lake, Texas, but by April 1899, it had closed down.[2]

Cullinan's construction crews completed their work in December 1898, and the first substantial refinery employing advanced technology in Texas was ready for operation. The installation consisted of a horizontal battery of four large boilerplate stills, twelve feet in diameter and thirty feet in length, each still with a 500-barrel capacity. These "shell" stills operated on the so-called batch principle: fuel-oil-heated lines under the stills "cooked" the oil, vaporizing singly the various components as the temperature increased. These gases then rose to the top of the still into a large, domed collecting outlet, moved out through coiled piping into a nearby condensing tank or "box," where water cooled the vapor into liquid form. The distilled components were then subjected to treating for the further removal of impurities. Since the major product of the refinery was to be illuminating oil, the construction of two large vertical agitator tanks, over sixty feet high, helped to render the distillate colorless and thus commercially appealing. An acidizing plant, through which the refinery runs also passed, removed or lowered the refined oil's sulphur content. The installation also included a two-story brick office building, a boiler plant, a brick warehouse and machine shop, several pumping stations, and a "tank farm" with fifteen additional storage units.[3]

At that time, a few eastern refineries employed more advanced

1. Dallas *Morning News,* July 24, 1898.

2. Warner, *Texas Oil and Gas,* pp. 15–16, 23.

3. The description of the Corsicana refinery is from "History of the Refining Department of the Magnolia Petroleum Company," unpublished manuscript prepared by the Publications Staff, Mobil Oil Company, Beaumont, Texas, pp. 10–12.

engineering practices, such as "continuous operation," steam distillation, and "cracking," or fractioning. These innovations came as refiners sought greater product diversification at lower manufacturing costs. Other petroleum manufacturing areas soon adopted these practices. The Corsicana refinery, for example, added two steam stills in 1901, which allowed for faster distillation at lower temperatures as steam was injected directly into the still. In 1908, the refinery converted to the continuous-operation process, which utilized gravity flow and controlled temperatures to manufacture several products during the same run. Yet, when the Corsicana refinery began operation in 1899, it was, in a technological sense, comparable to the usual "batch" operation of that period, which placed heavy emphasis upon the manufacturing of commercial grade kerosene, or illuminating oil.[4]

Cullinan's refinery began full operation in January 1899, running batches of 500 barrels of crude oil daily. In addition to kerosene, the plant also produced naphtha and quantities of better-grade, semi-refined fuel oil. The adaptability of local crude oil for commercial use was strikingly verified. Only the high-grade Appalachian field crude averaged more in kerosene yield (67 percent). Corsicana refinery runs produced 50 percent kerosene and 7 percent naphtha. The remaining 43 percent sold for use as fuel oil.[5] A quantity of gasoline, a by-product of the refining process for which there was then little market, was stored until spring rains came, when it was turned into swollen creeks and allowed to be swept away.[6]

The close relationship between the Corsicana partnership of Payne, Folger, and Cullinan and the Standard Oil organization was well-evidenced by the announcement, in February 1899, that Standard's marketing affiliate in the Southwest, the Waters-Pierce

4. *Ibid.*, p. 11. For further information on the various refining techniques of 1870–1900, see "Postwar Refining and Refining Organization," Chapter II in *The American Petroleum Industry: The Age of Illumination, 1859–1899*, by Harold F. Williamson and Arnold Daum (Evanston: Northwestern University Press, 1959).

5. U.S., Bureau of Corporations, *Report of the Commissioner of Corporations on the Petroleum Industry: Part I, Position of the Standard Oil Company in the Petroleum Industry* (Washington: U.S. Government Printing Office, 1907), p. 108.

6. "History of the Refinery Department of the Magnolia Petroleum Company," p. 11.

Company of St. Louis, would handle the refinery's products. Waters-Pierce quickly opened a local branch and transferred Aaron P. Robinson of St. Louis to Corsicana as general superintendent of that office. The company was reported to be rushing empty tank cars to Corsicana, as the first shipments of illuminating oil, marketed under the "Brilliant" and "Eupion" trademarks, were made on February 22, 1899, to Kerrville, San Marcos, Sherman, and Fort Worth, Texas. It was also pointedly announced by Waters-Pierce—certainly not unmindful of the need for a degree of acceptance at a time when a state antitrust suit was pending— that the local retail price of Corsicana-refined illuminating oil was to be at least two cents a gallon less than the retail price in other Texas cities.[7] Considering the obvious savings in transportation costs, this was a relatively easy promise for Waters-Pierce to fulfill. Investigations within two years revealed that, in both wholesale and retail prices, kerosene did sell from two cents to five cents a gallon cheaper in Corsicana than in other Texas cities.[8]

As would be expected, shipment of the first refined oil from Corsicana precipitated an outburst of civic pride, and each tank car bore a bunting banner boasting its contents as "Corsicana Refined Petroleum, Produce of the First and only Refinery in Texas or the Southwest!"[9] Pennsylvanian Joseph S. Cullinan joined in the spirit of the celebration and gallantly donated a thousand gallons of illuminating oil to the monument fund of the Navarro County Chapter, United Daughters of the Confederacy. The chapter designated Corsicana merchants Charles H. Allyn and Will M. Tatum to handle the sale of the oil, and local residents, eager to try a home-refined product and to aid a worthy cause, eventually bought all the contribution at a donative price of one dollar per gallon. The Daughters tendered Cullinan profuse thanks for his gesture, and, in a later meeting, they light-heartedly promised

7. "History of the Refining Department of the Magnolia Petroleum Company," p. 12; Dallas *Morning News*, February 22, 26, 1899.

8. U.S., Industrial Commission, *Report of the Industrial Commission on Trusts and Industrial Companies* (Washington: U.S. Government Printing Office, 1901), XIII, pt. 2, 820–821.

9. "History of the Refining Department of the Magnolia Petroleum Company," p. 12.

renewed and ingenious support in extolling the virtues of local petroleum. The Dallas *Morning News* wrote of the meeting:

A discussion on the utility of the Corsicana oil ensued. It was pronounced to be the most excellent for light, fuel, hair tonic, and complexion lotion.

The Daughters who have bald-headed husbands each ordered a gallon to be delivered at once. The maiden element of the chapter made secret resolution to test its qualities as a lotion. The whole chapter resolved not to hide its light under a bushel, so Mrs. Halbert and Miss Halbert were appointed to continue the correspondence with sister chapters concerning the disposition of our light at $1 per gallon.[10]

While such efforts were undoubtedly appreciated and helpful in promoting local acceptance of the refinery, the following months found the marketing arrangement with Waters-Pierce the most important factor in stimulating refinery output. After runs of 500 barrels of crude oil daily for the first months of operation, input increased to 600 barrels and then to 900 barrels; at the end of its first year, the refinery was operating at its full input capacity of 1,000 barrels of crude oil daily.[11] Because of the want of most records dealing with Cullinan's early activities at Corsicana, the total refined output for these years is unknown. Since Standard Oil considered the venture a personal enterprise of Folger and Payne, records of the Texas refinery production apparently were not kept by the combine.[12] Despite the refusal of the refinery to furnish any records, the United States Bureau of Corporations estimated that in one year, 1904, the refinery processed 190,000 barrels of illuminating oil. This was based on the Bureau's estimate that the refinery handled 80 percent of the Corsicana field's total crude oil production for that year, 455,281 barrels, which yielded a probable 50 percent of illuminating oil, or 190,000 barrels as measured in 50-gallon containers, then the standard shipping barrel for refined products.[13]

10. Dallas *Morning News*, March 5, 1899.

11. "History of the Refining Department of the Magnolia Petroleum Company," p.11.

12. Hidy and Hidy, *Pioneering in Big Business*, p. 415, n. 3; p. 770, n. 10.

13. *Report of the Commissioner of Corporations*, I, 266.

This production was unquestionably a welcome addition to
Waters-Pierce, which, between 1896 and 1904, controlled over
90 percent of the illuminating oil market in its Standard Oil-
assigned territory: Southern Missouri, Oklahoma Territory, Arkansas, western Louisiana, and Texas. Waters-Pierce, with the
exception of its Mexican operations, had no refining installations;
it purchased its manufactured products mainly from Standard
Oil's Whiting (Indiana) refinery. Even so, profits to the company
on each barrel of illuminating oil sold averaged 83 cents during
the period from 1895 to 1899, $1.32 from 1900 to 1904. Small
wonder that Waters-Pierce, with a market in excess of 90 percent
in its assigned territory, showed annual profits averaging 35.13
percent on total invested capital in the years from 1895 to 1904.[14]

While the Corsicana output could be easily absorbed into the
Waters-Pierce inventory, contributing to the firm's already substantial profits with savings in transportation and handling charges,
the refinery furnished only a minor part of the illuminating oil sold
by that company. Although Texas officials averred that as early as
1895 the company's sales in that state amounted to "half a million
barrels annually,"[15] later investigations showed the Waters-Pierce
sales of kerosene within its entire marketing territory were 585,228
barrels in 1902 and 627,260 barrels in 1903.[16] Obviously, the majority of Waters-Pierce sales inventory during Cullinan's Corsicana
years came from other sources within the Standard Oil combine.

The Corsicana refinery must be viewed in its proper perspec-
tive. Despite its vigorous management, which quickly brought the
plant to full operating capacity, the refinery remained, as initially
planned, a relatively small component within the Standard Oil
framework. It was planned small in order to permit the Standard

14. *Ibid.*, p. 19; Hidy and Hidy, *Pioneering in Big Business*, pp. 128, 514; U.S.,
Bureau of Corporations, *Report of the Commissioner of Corporations on the Pe-
troleum Industry: Part II, Prices and Profits* (Washington: U.S. Government Printing
Office, 1907), p. 47.

15. *State* v. *E. T. Hathaway*, 36 Texas Criminal Reports 261. This case was a criminal prosecution under antitrust laws of a Waters-Pierce manager, and it accompanied
the state's major civil suit against that company in 1896. Hathaway was found guilty
and fined $50, but the decision was reversed because of a faulty indictment by the
Texas Court of Criminal Appeals.

16. *Report of the Commissioner of Corporations*, II, 155.

executives, Payne and Folger, through limited participation, to experiment with Corsicana petroleum potential and to sample the political-legal atmosphere in a state where Waters-Pierce was already under antitrust attack. Ironically, even before these experiments could be completed, Corsicana, as a refining center, was doomed to a minor role by the geographical shift of the Texas petroleum industry to the lower Gulf Coast with the discovery of the Spindletop field in 1901.

One immediate result of the introduction of the refining phase at Corsicana, however, was the importation of experienced personnel to staff and supervise its installations. In the spring of 1898, as work on the refinery was about to begin, twenty-nine year old Edwy R. Brown arrived to assist Cullinan. Brown, a native of Ohio, had graduated from Marietta College and for the past four years had been employed by Standard Oil's Acme Refining Company in Olean, New York. Beginning as a common laborer at 15 cents an hour under the tutelage of William M. ("Uncle Billy") Irish, Acme's general manager and "The Schoolmaster" to a generation of future Standard Oil executives, Brown soon won promotion to stillman and, later, to assistant plant superintendent. He was called to New York, personally interviewed and selected by Folger for the Texas position. He became refinery superintendent under Cullinan; and after Cullinan had severed his Corsicana ties, Brown became president of the Corsicana and Navarro Refining companies. Later he was president of the Magnolia Petroleum Company and chairman of the Board of Directors of the Standard Oil Company of New York.[17]

Other "Olean graduates" who came to Corsicana in 1899 to assist Cullinan were Elmer E. Plumly and William H. Hastings. Plumly became superintendent of the Corsicana refinery in 1903, and was later vice-president and manager of all Magnolia Petroleum refineries with headquarters at Beaumont, Texas. Hastings also served later as a superintendent of the Corsicana refinery.[18]

17. "History of the Refining Department of the Magnolia Petroleum Company," pp. 8–9; *Magnolia Oil News (Founders' Number)*, April 1931, pp. 17, 24, 25.
18. "History of the Refining Department of the Magnolia Petroleum Company," p. 13.

To head the refinery's business office, Calvin N. Payne sent a young Pennsylvania accountant, W. C. Proctor. A former employee of the Marion Oil Company, a Standard Oil affiliate, Proctor was to become an officer in the Corsicana and Navarro Refining companies and, later, vice-president and treasurer of the Magnolia Petroleum Company.[19]

Cullinan was personally responsible for inducing another fellow-Pennsylvanian, William T. Cushing, to come to Corsicana in 1899. Cushing, an experienced pipeline builder, was placed in charge of constructing crude oil lines to connect the refinery with the field's gathering and storage system. Cushing was later to build pipelines for the Texas Company in the Beaumont area and, in 1919, he supervised construction of the major Ranger-Gulf Coast pipeline for the Humble Refining Company.[20]

These are but a few of the technical and administrative personnel migrating to Texas at the beginning of Corsicana's refining development. Although that area was not long to monopolize their energies, these experienced technicians and administrators were to establish a reservoir of talent for the subsequent development of the state's petroleum industry.

With a refinery in operation, the need for a continued and constant crude oil supply was obvious. Previously, Cullinan had assured this supply by contracts with the major producers specifying rates extremely favorable to his enterprise. Having no other reliable market available, local producers, with some complaint over low crude oil prices, nevertheless assented to periodic renewals of these agreements. Cullinan realized, however, that this situation might change in the near future. With exploration activity expanding the field's area and production, it was possible that other refineries might be attracted to Corsicana and compete with Cullinan for a substantial share of the field's output. Also, there was growing concern with the damage to the field's longevity in the haphazard drilling of numerous and ill-spaced wells by Corsicana's army of local and largely inexperienced operators. Cullinan, guided by

19. *Ibid.,* p. 10.
20. *Oil Investors' Journal,* February 20, 1910, p. 53; Larson and Porter, *History of the Humble Oil and Refining Company,* p. 145.

the experiences of his Standard Oil days, knew that these problems could be faced only by effecting further control over the field's exploratory and production activities.

From the beginning of his Corsicana association, the Pennsylvania entrepreneur had taken a keen oilman's interest in local exploration ventures. While his major efforts were devoted to his pipeline, storage, and refining installations, he occasionally took a partial interest with local businessmen in leasing prospective oil property. In the spring of 1898, Cullinan's participation in leasing activity increased as an old friend, Morris Egan, arrived from Washington, Pennsylvania. Egan, an experienced wildcatter, had some capital and, in partnership with Cullinan, he took up leases which were later drilled and found moderately productive.[21]

Ironically, it was in the tangled affairs of Corsicana's municipal water supply companies that Cullinan found an opportunity to make a major entry into exploration and production activity. Organization of the Corsicana Water Development Company, in 1894, led, by chance, to the initial discovery of petroleum at Corsicana when the company drilled an artesian water well.[22]

Meanwhile, an older firm, the Corsicana Water Supply Company, whose inadequacy as a water supplier encouraged the organization of the Water Development Company, had been thrust into financial difficulty by the advent of the new concern. To obtain a source of income to pay interest, at least, on bonded indebtedness, receivers of the older company had leased its plant, including distributing lines, several shallow wells, and an 800-acre parcel of land at the southeastern edge of Corsicana on which there was an earthen reservoir, to the new Water Development Company.[23]

By the fall of 1898, the Water Development Company was also having difficulty maintaining an adequate water supply for Corsicana. The heavy use of water in both the drilling and refining processes and the numerous well perforations made in the search for oil slowly dropped Corsicana's water table. The Water Develop-

21. Deposition of J. S. Cullinan, *Texas v. Magnolia Petroleum Company, et al.* (1913).

22. See Chapter 1, pp. 12–14.

23. Dallas *Morning News*, November 16, 1898.

ment Company drilled additional wells, but they were disappointing. The water was heavily mineral-laden and fit for only limited
industrial use. As a result, the Water Development Company was
forced to rely mainly on the wells and reservoir leased from the older
Water Supply Company. Litigation ensued, for the bondholders
of the older company claimed that this practice was a violation of
the lease agreement. Their property, they pointed out, was leased
as a supplemental water source. Its protracted use as a sole source
for the Water Development Company would soon deplete its value
and render the property worthless. The bondholders of the older
company thus asked that the lease be terminated. Attorney James
L. Autry, president of the Water Development Company, countered
with the argument that the lease was valid and that its provisions
should be enforced. After a prolonged trial, this case-in-equity was
decided by the United States Circuit Court at Fort Worth on
November 17, 1898. The decision was in favor of the Water Development Company and the lease was allowed to stand.[24]

The legal struggle had been closely followed by Corsicana oil
men with a concern paramount to their obvious need for an adequate water supply. This interest involved the 800-acre tract at
the southeast edge of Corsicana, property of the old water company
but under lease to the Water Development Company. The tract
was in the middle of the most productive area of the Corsicana
oil field and producing wells surrounded the property on all sides.
The itch which infected Corsicana oil men to drill this unexploited
property can well be imagined. It was reported that $10,000 had
been offered and refused for the oil rights on this tract of land.[25]
Meanwhile, the granting of these rights had been confused and
delayed by the litigation between the old and new water companies.

Cullinan was well aware of this tract's great potential and of the
considerable difficulty that would be encountered in gaining its
oil rights. Soon after his arrival in Corsicana, he had engaged the
law firm of McKie and Autry as counsel for his other Corsicana
enterprises. Through James L. Autry of the Water Development
Company, Cullinan learned further details concerning the tangled

24. *Ibid.*, October 8, November 18, 1898.
25. *Ibid.*, April 26, 1899.

affairs of the "Waterworks Property." A few months after litigation between the two water companies was settled, Cullinan, with the law firm's other partner, William J. McKie, quietly departed for New York City. Upon their return it was announced, to the surprise of the local business community, that Cullinan now owned the "Waterworks Property." From New York City banks, he had purchased a majority share of the old water company's $80,000 bonded indebtedness. This gave him a preferred claim on the assets of the company, and, in effect, he now controlled, or "owned," the company, including the much-desired tract of oil field land. Later, the purely water-producing assets of the company—the shallow wells, reservoirs, pumping stations, distributing lines, etc.—were sold for $15,000 to Autry's Water Development Company. Cullinan, of course, retained full mineral rights to the real estate of the old company.[26]

Although Calvin N. Payne guaranteed notes that made the acquisition of these valuable oil rights possible, the transaction represented largely an individual venture by Cullinan. McKie and Autry later admitted holding a minority interest in the waterworks tract, but their interest was probably assigned to them by Cullinan in payment for legal services. [27]

Exploratory drilling of the property was done under contract by the subsequently-organized Corsicana Petroleum Company, but Cullinan retained both royalty and production income. As predicted, the tract became very productive. The first well, drilled in October 1899, by the Corsicana Petroleum Company, initially produced 150 barrels daily. By instituting a policy of proper well spacing, Cullinan maintained the tract's production at a high and profitable level through succeeding years. During the month of September 1902, for example, the tract produced 3,981 barrels of crude oil. At that time, the price of Texas crude was generally depressed, following the Spindletop discovery; yet, the above production, sold to the Corsicana Refining Company at sixty-six cents a barrel, brought Cullinan income of $2,627.00 from this tract

26. *Ibid.*, April 26, 27, August 4, 1899.
27. W. J. McKie (Corsicana) to J. S. Cullinan (Beaumont), August 27, 1902, Cullinan Papers.

alone. Between 1899 and 1905, the tract was estimated to have produced over 400,000 barrels of crude oil.[28]

In acquiring this valuable property, Cullinan again demonstrated his ability to grasp business opportunity in a vigorous fashion. Also indicated is his realization that, as a newcomer to Texas, he would be well advised to share his opportunities with local professional and business interests. The sharing became, in effect, an exchange. Cullinan's opportunism, as evidenced by the role of McKie and Autry in the waterworks purchase, received competent legal guidance to insure its success and economic justification from a prestigious segment of the community. Local participants in turn, obviously received some monetary benefits from the association with Cullinan and, more importantly, perhaps, they held the hope that they could join in future business dealings with this experienced oil man. Local public opinion was thus conditioned to the acceptance of Cullinan's activities and the direct manner in which he marshaled his contacts and resources to achieve them. Even the Dallas *Morning News*, always distrustful of his Standard Oil connections, could not help admiring the Waterworks *coup*, stating that Cullinan's action was the "talk of the talent" and that the local impression was "that Cullinan has bought a great bargain."[29]

While the waterworks property was to prove very profitable for the Pennsylvania promoter, its acquisition by no means concluded his plans for further control of the field's production. In August 1899, James Garrity, a Corsicana banker and oil investor, and Cullinan returned from a business trip east. Cullinan then announced plans to organize a major production company by purchase and consolidation of oil interests held by smaller companies and operators. This venture, by corporate charter dated August 26, 1899, was organized as the Corsicana Petroleum Company. Its

28. Dallas *Morning News*, October 21, 1899; credit statement for September 1902, Corsicana Refining Company to J. S. Cullinan, October 10, 1902, Cullinan Papers; brief of W. J. McKie in Case No. 4862–A, *Matthew C. Cartwright et al.* v. *Corsicana Water Supply Company et al.*, July 1905, 13th Judicial District, Navarro County, Texas, in "Waterworks Property" Folder, James L. Autry Papers, Fondren Library, William Marsh Rice University, Houston, Texas. Hereinafter referred to as Autry Papers.

29. Dallas *Morning News*, April 26, 1898.

total authorized capital was $300,000, embodied in 3,000 shares of $100-par capital stock. The corporate purpose of the new company was limited soley to production activity: "said company organized for the purpose of owning and producing oil, gas, and other minerals." Cullinan was designated president and a director of the concern. Other directors were James Garrity, James E. Whiteselle, Garva E. Strong, and Calvin N. Payne.[30]

Payne, with his access to Standard Oil financial backing, was the major contributor and he held 2,500 of the company's 3,000 shares of capital stock. Cullinan and his Pennsylvania friend, Morris Egan, held 125 shares each, which were given in exchange for lease interests transferred to the new enterprise. The remaining 250 shares were divided among local business men who also assigned their oil properties to the new company: James Garrity, James E. Whiteselle, Joseph A. Edson, and Robinson M. Galbraith.[31]

The Corsicana Petroleum Company soon embarked upon an energetic program of acquiring further oil interests. Even before the charter of the company had been completed, Cullinan, acting as trustee, had arranged for the purchase and transfer to him of twenty producing properties. Although in most cases these transfers were made for an unstated "valuable consideration," the capital expended must have been sizable. In one transaction in which the purchase price was recorded, Cullinan, as trustee, paid $25,000 for five producing wells and mineral rights on several hundred acres of land owned by the Groesbeeck Cotton Oil, Gin and Compress Company.[32] By the end of August 1899, it was reported that the company already owned the production of thirty of the best wells in the field and had acquired mineral rights to "thousands of acres of potentially valuable oil land."[33]

A familiar problem, which had appeared in all the country's older oil fields, was again posed by the activities of Cullinan and his associates in organizing the Corsicana Petroleum Company. Was pressure applied upon the smaller operators to dispose of their oil

30. Copy of charter, Corsicana Petroleum Company, Texaco Archives, I, 38.

31. Deposition of J. S. Cullinan, *Texas* v. *Magnolia Petroleum Co. et al.* (1913).

32. Deed Records of Navarro County, CII, 530; CIII, 18.

33. Dallas *Morning News,* August 25, 1899.

properties to those who already controlled the field's storage and marketing functions and who could obviously impose economic retaliation if the smaller operator offered opposition? Some contemporary observers thought so—or feared so. The Dallas *Morning News*, reporting the organization of Corsicana Petroleum, stated that the new company initially "requested certain owners here to put a price on their oil well property." Later the newspaper was more explicit, reporting that "small operators have been under considerable pressure to sell" and that "the little fish will soon be digested by the big fellow. . . ."[34]

But if such pressure was at first exerted by the Cullinan group, it left no lasting heritage of ill will at Corsicana. Local opinion seemed to reflect the attitude of the same newspaper that feared undue pressure had been initially used. Upon second thought, the Dallas *Morning News* became resigned to the process of field consolidation, referring to it as an "inevitable drift . . . caused by old heads in the oil business. . . ." Cullinan's new enterprise, it was concluded, would stimulate the entire local petroleum industry, for it would mean "that the business of the whole might be economically conducted with more profit to those interested."[35]

There were few complaints about the price paid these smaller operators for their oil holdings. It must be assumed that the arrangements made were fair, even attractive, to local oil men and that they were glad to be relieved of the uncertainties and anxieties of producing operations. Since many of these operators were also landowners, they still retained a royalty interest in future mineral discoveries. This was a further incentive for the smaller oil producers to relieve themselves of the expense and uncertainty of present operations, but with a guarantee to share substantially, without capital outlay, in any future production found on their properties.

Participation in the Corsicana Petroleum Company by respected Corsicana businessmen, such as the banker, James Garrity, and the former mayor, James Whiteselle, again gave a venture largely financed and directed by outsiders the flavor of local support and approval. Moreover, the ever-present concern that the ties between the

34. *Ibid.*, August 15, 25, 1899.
35. *Ibid.*, August 2, 1899.

Corsicana ventures and Standard Oil's litigation-ridden affiliate, Waters-Pierce, might stimulate further public and legal disapproval surely influenced and guided Cullinan and his associates in these transactions. The conclusion was inescapable: most of the lesser Corsicana operators were pleased to be out of the oil business at a fair price and, for those holding royalty interests, with the prospect of further production income without additional capital expenditure.

The Corsicana Petroleum Company, by purchase of existing production and exploration of leased property, quickly became one of the major producing companies in the field. While production for the years from 1899 to 1902 is unknown, incomplete records maintained by Standard Oil affiliates indicate the company still produced 46,000 barrels of crude oil in 1903, 37,000 in 1904, 29,000 in 1905, and 38,000 in 1906.[36] For Cullinan and the other stockholders, it was unquestionably a profitable venture. The company paid an annual dividend equivalent to 12 percent of its capital in the years 1901 through 1904, and in addition it paid special dividends totaling 25 percent of its capital through the same period.[37] During subsequent years, the company continued its successful operations in the Corsicana field and, in 1911, pioneered in the development of the major Electra field of north Texas. It also obtained a permit to do business in Oklahoma in 1912 and operated leases in the Glen Pool area. In 1916, after acquisition of a controlling interest of its stock in 1911 by the Magnolia Petroleum Company, Corsicana Petroleum was absorbed into that company as its "Producing Division."[38] Cullinan understandably continued to hold his profitable stock interest in this company even after he was no longer active in its management. But in 1910, fearing a conflict of interest while serving as president of the Texas Company, he traded his 125 shares to Calvin N. Payne for real estate holdings in Corsicana.[39]

36. Hidy and Hidy, *Pioneering in Big Business*, p. 375.

37. W. C. Proctor (Corsicana) to J. S. Cullinan (Beaumont), April 22, 1902; C. N. Payne (Oil City, Pa.) to J. S. Cullinan (Corsicana), November 4, 1902; J. S. Cullinan (Beaumont) to David Iseman (Washington, Pa.), November 25, 1904, Cullinan Papers.

38. Rister, *Oil! Titan of the Southwest*, p. 113; *Oil and Gas Journal*, June 19, 1913, p. 12; "History of the Refining Department of the Magnolia Petroleum Company," p. 14.

39. Deposition of J. S. Cullinan, *Texas v. Magnolia Petroleum Co. et al.* (1913).

At the same time that Cullinan was utilizing outside capital to organize a major producing company and to control a substantial portion of the field's output, out-of-state capital was also acquiring another large Corsicana producer, the Southern Oil Company. This company, formed in October 1897 was a successor to the Corsicana Oil Development Company, the pioneer in the field's exploration. Following the exodus of the wildcatters Galey and Guffey, the incorporation of Southern Oil by local businessmen was enthusiastically hailed as evidence that Corsicana's petroleum development was to be guided by "home" interests.

Unprophetic and ironic as these assertions later became, Southern Oil seemed in the beginning to prosper under local management and capital. During 1898, the company produced 120,360 barrels of crude oil and distributed $40,000 in dividends to the following local stockholders and directors: Ralph Beaton, president and general manager; H. G. Damon, Aaron Ferguson, Howard W. White, Fred Fleming, Allison Templeton and S. W. Johnson. At the end of 1898, Southern Oil owned forty-four producing wells and had leased exploration rights to over 30,000 acres of land.[40]

While this made Southern Oil the largest crude oil producer in the field, the company soon faced major problems. As field development became complex and demanded further capitalization, Southern Oil was penalized by inexperienced management which seemed determined to milk the company of immediate profits and which gave little thought to the needs of business expansion. Although incorporated for an authorized capital of $300,000, nothing close to this amount had actually been paid in to the enterprise. By unanimous consent of the seven stockholders a nominal assessment of $5 a share, on $100-par value stock, had been made at incorporation.[41] As the company needed further funds for field equipment and drilling ventures, its managers preferred to raise capital by bank borrowing rather than by a further call on capital stock or retention of operating income.

40. Meeting, January 4, 1899, "Minutes, Board of Directors' Meetings, 1897–1901," p. 32, Southern Oil Records, Warner Collection. Records of the company from this collection are hereinafter referred to as SOR.

41. Meeting, November 1, 1897, "Minutes," p. 5, SOR.

This policy soon had the company heavily in debt. At the end of 1898, it had three mortgage notes totaling $51,000, at 10 percent interest, with the Corsicana banking house of Fleming and Templeton, whose partners, Fred Fleming and Allison Templeton, were also stockholders and directors of the company. An inventory of the company's assets appraised in May of that year totaled only $28,102.36. Yet the company paid dividends totaling $20,000 from operating income earned, January 1 to June 30, 1898, and the directors declared an additional $20,000 in dividends to be paid from earnings during the last six months of that year. The latter dividend was not actually paid to the stockholders, for bankers Fleming and Templeton understandably demanded assignment of these dividends to them in reduction of the company's indebtedness.[42]

Corsicana business circles were not greatly surprised, therefore, when it was announced, in April 1899, that the company had been sold to a syndicate of St. Louis industrialists headed by street railway magnate Henry T. Kent. The company was purchased for $250,000 in cash, which was the amount asked by the local owners from the beginning of their negotiations with the St. Louis group. Corsicana oil men felt that this was a fair price. The former owners, even after settlement of indebtedness, realized a substantial profit, and, it can be imagined, also gave a collective sigh of relief to be out of the oil business.[43]

Local operators further welcomed the sale because it was at first rumored that Kent's group intended to construct a refinery at Corsicana. They reasoned that the competition for field output between the new refinery and Cullinan's enterprise would raise crude oil prices. Cullinan was eagerly interviewed for his reaction to this rumor. But he was "calm and unruffled," and it was concluded that Cullinan, well aware of the St. Louis group's plans, had no fear of such competition developing. Within a few days, Henry T. Kent affirmed that the new Southern Oil management had no intention of building a competing pipeline or refinery at Corsicana. Rumors to the contrary were "pure imagination," since the company would

42. Meetings of May 6, June 6, October 3, 30, 1898, "Minutes," pp. 13, 20, 29, SOR.
43. Meetings, April 1, 15, 1899, "Minutes," pp. 33, 37, SOR; Dallas *Morning News*, April 17, 1899.

continue with renewed emphasis, its exploration and production activities.[44]

Given assurances that the new management of Southern Oil did not intend to compete with his gathering and refining enterprises, Joseph S. Cullinan unquestionably welcomed a revitalized producing company in the Corsicana field. Since his refinery remained the major market for crude oil, such a company could further guarantee a dependable petroleum supply at advantageous prices. The new directors of Southern Oil immediately informed Cullinan that they would offer him the company's entire output and, during the next two years, the refinery purchased an estimated 90 percent of the company's crude oil production.[45]

The new owners also completely reorganized the management of Southern Oil. Henry M. Ernst, an experienced oil man from Olean, New York, was hired as the company's general manager. Ernst had been active in Pennsylvania production ventures since the late 1870s and, in 1885, he organized the Lima Drilling Company, a pioneer producer in the Ohio petroleum field. As that concern was later merged into the Ohio Oil Company, Ernst became the latter company's president. He resigned from Ohio Oil in 1899, however, when Standard Oil acquired control and he had operated as an independent producer in the Olean area until coming to Corsicana. The directors attracted him to Texas with a $4,000 annual salary and, more important, with the offer of 500 shares of Southern Oil's preferred stock. Later an official of the Houston Oil Company, Ernst was another example of experienced oil talent lured to Texas by the Corsicana field and whose presence facilitated the future development of the state's petroleum industry.[46]

The new directors of Southern Oil next made provision for a substantial increase in the company's capitalization. The total capital stock was increased from $300,000 to $600,000 and additional capital was authorized through the issuance of $500,000 preferred,

44. Dallas *Morning News*, April 6, 17, 1899.

45. Meeting, April 15, 1899, "Minutes," p. 37, SOR; "Prospectus, Houston Oil Company of Texas, 1901," Houston Oil Company Records, Warner Collection. Records of this company from this collection are hereinafter referred to as HOCO.

46. *Derrick's Hand-Book*, I, 958; Meeting, November 8, 1899, November 6, 1900, "Minutes," pp. 48, 54, SOR.

"Six Percent, Cumulative," stock. While the capital, or common, stock of the company continued to represent only a nominal paid-in contribution, the directors raised expansion capital through periodic calls on the preferred stockholders. Between May 1899 and April 1901, periodic calls totaling $65 per share had been made on the preferred stockholders.[47]

This increased working capital was soon put to use. The company also began the purchase of existing production and lease rights from the field's smaller operators. In May 1899, shortly after the St. Louis group assumed control, Southern Oil acquired forty additional wells from the Hardy Co-Operative Oil Company, operator Rod Oliver, and the partnership of Oliver and Stribling. The company also secured oil rights on 5,000 acres of land previously leased by the Texas Petroleum Company. In these transactions, Southern Oil paid for the property by tendering $66,313.51, cash, and $15,366.27 in short-term, ninety-day-maturity notes.[48]

Moreover, the new management of the company showed its willingness to expend sizable amounts in the search for new production. During 1898, the company, under Corsicana ownership, had drilled nineteen new wells. In 1899, with new management and capital, the company drilled fifty-five wells, obtaining production from thirty-five of them. The next year, 1900, this pace was maintained, with fifty-six new wells drilled, forty-four of which were producers.[49] With the advent of the St. Louis group insuring adequate capital and experienced management, the Southern Oil Company literally gained a "second wind," which reaffirmed its position as the Corsicana field's major production concern.

In July 1901, the stockholders of Southern Oil decided to sell the company's assets, which then included 115 producing wells and 200,000 acres under lease, to the newly-organized Houston Oil Company of Texas for $1,000,000. A detailed examination made at the time indicated that operations under St. Louis capital had been very profitable. During the two years from May 1899 to May

47. Meetings, November 8, 1899, April 30, 1901, "Minutes," pp. 48, 56, SOR.
48. Meeting, May 6, 1899, "Minutes," p. 39, SOR.
49. This summary of drilling activity was compiled from the "Well Record and Log Book, 1897–1904," SOR.

1901, the company averaged annual gross earnings of $150,380.64, less average annual operating and general expenses of $32,692.51, which resulted in average annual net earnings of $117,688.13. The company's crude oil production during that period averaged over 529 barrels daily or about 194,000 barrels annually.[50]

The decision of the St. Louis owners to sell the company's assets by no means indicated a disenchantment with Corsicana operations. Rather, the owners of Southern Oil were given an opportunity to use their property as a basis for capital participation in an even more ambitious enterprise, the Houston Oil Company of Texas. In 1900, New York financier Patrick Calhoun, later a co-promoter of the Houston Oil Company, became a heavy investor in the Southern Oil Company. Upon the 1901 organization of Houston Oil, which held title to over 800,000 acres of east Texas timberland near the Beaumont-Spindletop petroleum field, Calhoun persuaded his fellow Southern Oil investors to sell, or trade, their property for a $1,000,000 stock interest in his new enterprise.[51] This sale gave Houston Oil immediate crude oil production from the Corsicana properties and afforded the St. Louis group a chance to participate in what seemed to be a very promising project. It was a transaction, however, that the former Southern Oil owners undoubtedly regretted. The early years of the Houston Oil Company were filled with acrimony between Patrick Calhoun and his co-promoter, Texan John Henry Kirby, resulting in prolonged litigation and receivership. It was almost two decades before the company fulfilled its promise and became a profitable producing concern. This development began in 1916, when a company headed by Joe Cullinan assumed direction of petroleum activities on Houston Oil Company lands.

Cullinan's key role in the organization and managerial direction of a major producing unit, Corsicana Petroleum, and a manufacturing plant, Corsicana Refining, thus signaled the initiation of

50. "Report of Haskins and Sells, Certified Public Accountants, August 10, 1901," HOCO.

51. For further information on the role of the Southern Oil properties in the formation of the Houston Oil Company, see John O. King, *The Early History of the Houston Oil Company of Texas, 1901–1908* (Houston: Texas Gulf Coast Historical Association, 1959), pp. 18, 27–28, 30.

integrated operations within the Texas petroleum industry. More-over, the resulting field stability enticed additional out-of-state capital to purchase the field's largest producing company, Southern Oil. Because of his control of the field's market, this revitalized unit offered no competition but instead served as a valuable complement to Cullinan's operations.

4. Cullinan at Corsicana:
Legislation, Change, and Conclusion

THE DEVELOPMENT of the Corsicana oil field soon demanded another pattern that significantly influenced the future of Texas petroleum development: the role and responsibility of public policy, as expressed in legislative fiat, to compel field conservation practices. Again, Texans with no previous knowledge of such legislation would look to Joe Cullinan and his wide experience in oil-producing states with petroleum conservation laws for leadership and advice.

The need for the initiation of this type of legislation had been obvious to Cullinan from the first time he inspected the field; for, in the early years of development at Corsicana, before Cullinan had brought a high degree of stability through control of marketing and transportation functions, inexperienced local operators had, often willfully, committed a variety of conservation sins. They had crowded too many wells into productive areas, wasted crude oil through inadequate storage facilities, and "flared" away valuable pockets of natural gas. But Cullinan and other experienced oil men found most appalling the lack of local concern with the proper "plugging" of abandoned wells. Even the then relatively rude study of reservoir engineering had demonstrated that a few unplugged or carelessly-plugged abandoned wells soon had an adverse effect on field production and longevity. Left open, these wells soon filled with surface water which ran through to the oil-bearing strata, diminishing production of neighboring wells and shortening the life of the entire field. As a result, states with prior petroleum development had quickly resorted to legislation to enforce proper abandonment procedures. Pennsylvania, with the nation's first commercial fields, pioneered in petroleum conservation legislation with a compulsory well-plugging law enacted in 1878. New York enacted similar legislation in 1879, Ohio in 1883, and West Virginia in 1891.[1]

To Cullinan, the Corsicana situation was intolerable. "In this country," he later wrote, "such legislation was essential where there were so many small holders, many of them wholly inexperienced [and] . . . not likely to go to the expense of plugging wells unless compelled to do so."[2] He then organized a series of informal meetings with Corsicana business and civic leaders, many of whom were oil field investors, and soon persuaded them to sponsor such legislation.[3]

On February 14, 1899, the Navarro County delegate to the Texas House of Representatives, Robert E. Prince, introduced House Bill No. 542, a measure "to regulate the drilling, operation, and abandonment of gas, oil, and mineral water wells and to prevent certain abuses connected therewith."[4] The bill was then referred to the House Committee on Mining and Minerals. Culli-

1. Walter L. Summers, "The Modern Theory and Practical Application of Statutes for the Conservation of Oil and Gas," *Legal History of Conservation of Oil and Gas: A Symposium* (Chicago: American Bar Association, 1939), p. 1.

2. J. S. Cullinan (Beaumont) to W. J. McKie (Corsicana), February 1, 1905, Cullinan Papers.

3. *Ibid.*

4. Texas, *House Journal, 26th Legislature, 1899* (Austin: State Printers, 1900), p. 404.

nan, with Whiteselle and Garrity, Corsicana businessmen and oil investors, subsequently appeared before that committee in support of the bill.[5] Favorably reported by the committee on February 25, the bill passed its first and second readings in the House on March 6 by an overwhelming margin, 92 yeas, 1 nay.[6] This margin of more than a two-thirds majority allowed suspension of the rule requiring a third reading and the bill was passed by the House and sent to the Senate with an appended explanation that "as surface water in wells now abandoned is calculated to ruin the oil field in Navarro County, an emergency is created. . . ."[7] The Senate, as recommended, acted quickly: the bill was favorably reported by the Committee on Mining and Irrigation on March 15, and passed in final form on March 23 (yeas, 25; nays, 1).[8] Texas' first petroleum conservation statute became effective immediately on March 29, 1899, as Governor Joseph D. Sayers signed the bill.[9]

The new law specified in detail the manner in which abandoned wells were to be plugged. Section Two stated that the well casing was to be securely filled with rock, sediment, or a mortar mixture composed of two parts sand and one part cement. Sufficient material was to be packed into the casing to fill it to the depth of two hundred feet above the top of the first oil-bearing rock, and in such a manner as to prevent gas and oil from escaping. It was the responsibility of both the well operator and the landowner to comply with this procedure. If they failed, then it would be lawful for any person, after written demand to the operator and the landowner, to enter the premises and perform the well-plugging procedures. The land owner and well operator were then liable for the plugging cost and, if it was not paid, the amount became a lien on the equipment of the operator and the title of the landowner.

While the section dealing with well-plugging was viewed as the

5. Dallas *Morning News*, February 23, 1899.

6. Texas, *House Journal, 26th Legislature, 1899*, pp. 661–662.

7. *Ibid.*

8. Texas, *Journal of the Senate, 26th Legislature, 1899* (Austin: State Printers, 1900), pp. 432, 452, 534.

9. Texas, *House Journal, 26th Legislature, 1899*, p. 875. For a full text of the statute, see H. P. N. Gammel, compiler, *Laws of Texas, 1897–1902* (Austin: Gammel Book Company, 1902), p. 68. The following discussion of the various sections of this law refers to this source.

most essential part of the new law, the statute prohibited several other blatant oil field abuses. Section One, for example, specified that wrought iron or steel casing was to be required in drilling operations to exclude all surface or fresh water from penetrating the oil or gas-bearing rock. Section Three prohibited the waste of natural gas by requiring all gas wells to be confined within ten days after discovery. Section Four sought to prevent further wastage of natural gas by prohibiting indiscriminate "flaring" or burning about the well site for illuminating purposes. A permissible type of enclosed burner ("Jumbo") was specified and the flaring of gas in any type of burner during the daylight hours was prohibited. In conclusion, Section Five imposed a penalty of $100 for violation of the above provisions which was recoverable, plus costs of suit, in the county courts where the offense was committed. The penalty, when collected, was to be divided equally between the school fund of the county where the suit was brought and the individuals initiating the suit.

Cullinan was pleased that the state legislature had acted with alacrity and he generally approved of the new legislation. He was disappointed, however, that the legislature failed to incorporate into the new law several major suggestions he had made previously at the hearings in Austin. He had testified that the penalties to be assessed were too light and, more important, that the proposed statute should designate a public official to see that the well-plugging procedures were complied with. Cullinan felt that such an official should be empowered to make property inspections, to require a log of each well drilled, and a sworn statement attesting to the proper plugging of abandoned wells. But the legislature had not seen fit to incorporate these provisions in the 1899 act; Cullinan, in his disappointment, prophesied that eventually the legislature would acknowledge its mistake and amend the law to include his suggestions.[10]

His prophecy was correct. In 1905, following deterioration of conservation practices in the Gulf Coast fields, the Texas legislature amended the 1899 statute. Penalties for violation of statute

10. Cullinan (Beaumont) to W. J. McKie (Corsicana), February 1, 1905, Cullinan Papers.

provisions were increased, to range from $500 to $5,000. County courts were given injunctive powers and authority to appoint an oil field superintendent to enforce conservation laws.[11] In 1917, after further legislation empowered the State Railroad Commission to enforce conservation measures, that body soon required well records similar to those proposed over a decade earlier by Cullinan.[12]

Thus the initial relationship in Texas between petroleum conservation and compulsory legislation was established in the development of the Corsicana field. A pattern was invoked that would guide the future of oil and gas conservation within the state. The early Texas petroleum industry, as evidenced by the Corsicana exploitation, was composed of eager, opportunistic men of diverse experience and temperament. Compulsory legislation, as demanded by Joseph Cullinan, the industry's most experienced leader in Texas at the time, became the major method of enforcing petroleum conservation.

Social disorder and change also came to Corsicana with the discovery of oil and the subsequent development of the industry. Although these events were apparently not as sudden, dramatic, or well publicized as later discoveries, Corsicana was nonetheless Texas' first oil-boom town. A small locality, essentially agrarian in economic and social values, it was introduced to the problems and stresses brought by a vigorous and largely alien industrial order. It was ironic that on July 24, 1894, as drillers struggled to complete the artesian well that launched the Texas petroleum industry, the Dallas *Morning News* found the most noteworthy recent happening in the area to be the news that a valuable mule was struck and killed by a freight train near Corsicana. But then oil came to this bucolic cotton town. The pace of local events and the style of news reporting quickly changed.

The first intimation that the discovery of petroleum meant notoriety occurred as the oil fraternity rushed to Corsicana to in-

11. Texas, *General Laws of the State of Texas, Twenty-Ninth Legislature, Regular Session, 1905* (Austin: State Printing Company, 1905), p. 228; Warner, *Texas Oil and Gas*, p. 49.

12. Warner, *Texas Oil and Gas*, pp. 57, 60–63.

vestigate the field. Scores of oil scouts, drillers, and potential investors drifted in and out of town, their number continually increasing as the news of the field's potential circulated. Corsicana's three hotels, the Commercial, Mallory, and Collins, soon filled to capacity and a transient and weary oil man counted himself among the fortunate with a cot in a hotel corridor or even a chair in the hotel lobby.[13]

The oil excitement brought other outsiders of lesser virtue who also apparently sought opportunity. Corsicana residents soon complained about the great number of "tramps" and "roving characters" infesting the town and attributed a rash of pilferage and burglary, even a great number of clothes line thefts, to this invasion. Oil excitement, as many other Texas communities adjacent to petroleum fields were later to learn, spawned the rapid proliferation of saloons, gaming, and bawdyhouses. Corsicana once again led the way. Activity near the Houston and Central Railroad tracks, particularly on paydays, was brisk and carried on at all hours "with little restraint on noise or nonsense." Periodic reports of violence and mayhem emanating from this section were not unusual. A typical incident at a local saloon involved a recently arrived Pennsylvania pipe-gauger, "named French," who was severely battered with a brick by a Corsicana resident, "one Hardy Bowles." Finally, city officials did close a few of the most notorious establishments and decreed that the remainder strictly observe city ordinances requiring suspension of business activity on Sundays.[14]

But as the field developed, technical and administrative personnel, often migrating to Corsicana with their families, added a more stable element to the community. An immediate result of this migration was a housing shortage that quickly and sizably raised rents. The Dallas Morning News reported that dwellings worth only a few hundred dollars rented for $10 to $12.50 a month. This shortage of housing eventually brought a spirited boom in residential construction. During 1899, 350 new buildings were constructed in Corsicana, 80 percent of which were residences. This was twice

13. Dallas *Morning News,* December 9, 1897, January 23, 1898.
14. *Ibid.,* February 4, November 23, 1898, October 8, 1899; "History of the Refining Department of the Magnolia Petroleum Company," p. 13.

the number of houses built in any previous year and, since most of the new homes were substantial residences, ranging in cost from $3,000 to $6,500 each, this represented a total construction expenditure three times that of any previous year. The report also noted that, during the same year, six office buildings were constructed in Corsicana at a total cost of $110,000.[15]

The wages paid to employees of the petroleum industry further stimulated Corsicana's economy. In 1898, at the height of field development, it was estimated that 100 men were employed in drilling operations. Wages paid for this work ranged from $1.30 to $2.50 for a twelve-hour day. The field collecting and storage system employed about forty additional workers. Most of them were skilled, having been imported by Cullinan from other oil regions. Wages ranged from $1.80 per ten-hour day for pumpmen and boiler-makers to $3.00 a day for foremen and supervisory personnel. Later, the refinery employed about fifty workers and wages ranged from $1.50 per twelve-hour day for laborers to $2.00 for stillmen, pipe fitters, etc., on to $3.00 for yard foremen and chief stillmen. While these wages were lower than those paid by Standard Oil companies along the eastern coast, they were comparable to wages paid in the midwest petroleum regions (West Virginia through Indiana) where wages were 15 to 20 percent lower than "Eastern" pay scales.[16]

Corsicana businessmen obviously welcomed this economic boost and were disappointed that there was not more of it. They had gained the impression, as Cullinan's refinery was under construction, that the long-awaited installation would employ hundreds of men. When the refinery began operation with less than fifty employees, there was some disillusionment. "The refinery," a newspaper reported, "does not give employment to so large a number as is generally supposed," but it added hopefully that "the less than fifty men [employed] are all paid good wages."[17] This belief that the development of the local petroleum industry automatically

15. Dallas *Morning News,* October 2, 1898, December 4, 1899.

16. *Ibid.,* September 12, 1898; "History of the Refining Department of the Magnolia Petroleum Company," pp. 11, 13; Hidy and Hidy, *Pioneering in Big Business,* p. 589.

17. Dallas *Morning News,* December 14, 1898.

meant a sizable influx of permanent residents persisted among the local citizens. In December 1898, the city council estimated Corsicana's population to be in excess of 11,000 and authorized a special census. If this census verified the town's population to be more than 10,000, the council could then petition the state legislature for a home rule charter. But much to the disappointment of the city fathers, and to their puzzlement as they surveyed the bustling oil activity about them, the special census indicated that the town's permanent population was still under 10,000.[18]

Nevertheless, the oil excitement brought people to Corsicana whose backgrounds and attitudes often differed from those of the older inhabitants. In the mingling of these groups, in the shaping of local institutions to meet the new challenges which arose, the oil industry brought to Corsicana a further degree of cultural diversity and civic responsibility. For instance, in May 1899, Samuel M. ("Golden Rule") Jones, oil equipment manufacturer and mayor of Toledo, Ohio, inspected the Corsicana oil field. Before leaving town, he was persuaded to give an address on "The Social Problem" at Corsicana's Opera House. Although the text of the speech was not reported, the audience was characteristically described as "large and enthusiastic."[19]

Since many of the incoming oil field workers were single, or married men temporarily absent from families, local churches and fraternal organizations did what they could to relieve boredom and counteract the temptations of Corsicana's gaudier sections by providing recreational opportunity and religious solace. Local members of the YMCA established canteens at various locations throughout the oil field and brought Sunday services to drilling crews whose round-the-clock schedule prevented church attendance. The local Catholic priest, Father Michael J. Kelly, announced his intention to solicit funds to open an "Oil Man's Club" where off-duty workers could lounge, read, and write letters home. Father Kelly insisted that his project was non-sectarian and open to all regardless of creed. This avowal did little to allay the fears of sev-

18. *Ibid.*, January 5, 9, 1899. The 1900 U.S. Census gave Corsicana's population as 9,313.

19. Dallas *Morning News*, May 24, 1899.

eral Protestant ministers, who were well aware that many of the recently-arrived oil field workers were Irish Catholics. The ministers apparently felt Father Kelly's club would serve predominantly as a Catholic haven and a threat to proselyte the unsuspecting. Led by the Rev. Amos B. Ingram, these ministers published in a local paper a note warning Protestants not to encourage the project.[20]

Here was a situation that enraged the Irish pride of Joe Cullinan. He came from a family rooted deep in the Irish tradition, whose members were devout communicants of the Catholic faith. Although Cullinan throughout his life disclaimed membership in any church, he remained very sensitive to anti-Catholic prejudices— even more so, of course, to the slightest suggestion of bias toward Irish Catholics. He must have quickly steeled Father Kelley in his club plans, for despite some local Protestant condemnation, it was significantly reported a few days later that the priest addressed Cullinan's refinery workers, seeking contributions for the project. With this show of open support, Father Kelley was soon able to obtain contributions from both townspeople and oil workers. The Oil Man's Club, located in rented quarters above Freedman's Store on Beaton Street, opened in January 1899, with sixty members.[21]

The petroleum industry led directly to the development and expansion of other Corsicana facilities. In November 1897, the locally owned Mutual Telephone Company was sold to the Southwestern Telephone and Telegraph system. Directors of the local company gave as a major consideration in the sale the fact that, with its growing petroleum development, Corsicana should have linkage with long-distance lines. The Western Union Telegraph Company opened a new office; and Henry T. Kent, leader of the St. Louis group which purchased the Southern Oil Company, obtained a franchise to construct an electric street railway. A number of oil field accidents involving transient workers pointed up the need for a local hospital. Corsicana business leaders organized a committee, on which J. S. Cullinan served, to investigate the possibility of establishing the first public hospital in Navarro County. Although the committee reported affirmatively on the need, how-

20. *Ibid.*, November 30, December 19, 1898.
21. *Ibid.*, December 23, 1898, January 17, 1899.

ever, it was to be a few years later, 1908, until the first hospital was established in Corsicana.[22]

Another result of the petroleum development at Corsicana was the introduction of local residents to careers in the oil industry. Examination has been made of the experienced oil men who came to Corsicana to guide this development. But it was equally significant, particularly to the future of Texas petroleum, that many local residents, stimulated by the excitement and opportunity of a home-town oil boom, subsequently sought professional and administrative careers with the industry.

An excellent example is the career of James L. Autry, local attorney and civic leader, whose firm was legal counsel for Cullinan's Corsicana enterprises. Autry was born at Holly Springs, Mississippi, on November 4, 1859. His father, later a colonel in the Confederate Army, was killed at the Battle of Stones River, December 31, 1862. Young Autry came to Texas in 1876 to manage family lands in Navarro County granted to his grandfather, who had served in the Texas revolutionary army. He settled in Corsicana, read law with a local attorney, was admitted to the state bar in 1880, and eventually served two terms as a county judge before forming a law partnership with W. J. McKie. Autry married Allie Kinsloe, daughter of a prominent Navarro County planter family, in 1896.[23]

The close friendship first established between Cullinan and Autry at Corsicana continued for many years. Autry went with Cullinan to Beaumont in 1902 and became general attorney for and a director of the Texas Company. When Cullinan resigned from that company in 1913, Autry also severed his association, joining Cullinan in the formation of the Farmers Petroleum Company and, later, the American Republics Corporation. At his death in Houston on September 29, 1920, Autry was a widely recognized authority on all phases of petroleum law.

Younger men, natives of Corsicana, who began long and prominent careers in the Texas petroleum industry, were Guy Carroll,

22. *Ibid.*, November 12, 1897, November 30, 1898, May 18, 1899; Taylor, *History of Navarro County*, p. 127.

23. Ellis A. Davis and Edwin H. Grobe, editors and compilers, *The New Encyclopedia of Texas* (Dallas: Texas Development Bureau, n.d. [circa 1925]), I, 1011.

Ernest Carroll, Eric H. Buckner, and Carlton D. Speed, Jr. The brothers Carroll, sons of a local lumber dealer, were both employed by Cullinan's refinery following graduation from Corsicana High School. Later, they accepted positions with the Texas Company and eventually became officers in that organization. Buckner started to work in 1897 at the age of seventeen for the J. S. Cullinan Pipe Line Company. He later moved to Beaumont during the Spindletop boom, and still later, in 1922, became president of the Houston Oil Company of Texas.[24] Speed, son of a drilling contractor who drilled a producing well at Corsicana in 1898, is an example of the second-generation careers in petroleum stimulated by the field. He was born there in 1903 and literally grew up in the local oil field. His earliest recollection was of a drilling rig grinding away on a town lot in Corsicana just a few feet from the family home. Later, he was given the chore of tending the small one-cylinder gasoline engines used to pump nearby wells. With such a background, young Speed later pursued studies at Texas A. & M. University and the University of Chicago in geology and petroleum engineering. At the time of his death in Houston on January 17, 1970, Speed was recognized as one of the most prominent of the nation's petroleum consultants and independent producers, having discovered and developed some ten Texas fields during his career.[25]

Industrialization brings changes, economic and social, and the robust impact of the petroleum development on Corsicana served as a harbinger of even more pronounced change in later twentieth-century Texas. Yet, as these changes occurred, they were often intangible and difficult even for the participants to describe and assess. Perhaps, in the final summation, the most discernible change apparent at Corsicana was the introduction to Texas of the oil-boom state of mind. A placid, rural type of economy and social order was initially conditioned to accept the risks of industrial opportunity, which were to be offered, even more dramatically, by the state's further petroleum development.

24. For biographical sketch of Guy Carroll, see Davis and Grobe, *New Encyclopedia of Texas*, II, 1633; for Ernest Carroll, *ibid.*, I, 369; for Eric H. Buckner, *ibid.*, I, 378.

25. Interview with Carlton D. Speed, Jr., Houston, Texas, October 15, 1962; Houston *Chronicle*, January 19, 1970.

Corsicana's significant contribution, unfortunately, has been overshadowed by the discovery of the prolific Gulf Coast fields in 1901, which soon drew the major portion of the state's petroleum industry to that area. Had these fields remained undiscovered, however, the industry at Corsicana would have continued as permanent and productive. In the succeeding years, the Corsicana oil field was greatly expanded through additional discoveries. The Powell district, ten miles southeast of the original field, was developed in 1905. The next year, this new pool, combined with the older field, produced 1,005,843 barrels of crude oil—a peak year for shallow-depth production in the Corsicana area. Output leveled off to an average of 508,000 barrels annually during the next fourteen years but, in 1920, flush production came, thirty miles to the south, with the discovery of the Mexia field, which in 1923 reached a peak output of 33,937,513 barrels. That same year, the Powell area became the most productive field in Texas with the discovery of a deeper oil sand, at 2,963 feet, underlying the older producing strata; and, ironically, some of the best wells in this development were on land owned by Cullinan's old Corsicana associate, W. J. McKie. Output from this new Powell area totaled 84,777,076 barrels during the first three years of its existence. Even the original shallow-depth field within the Corsicana town limits has retained production some seventy years after its early development. As late as 1954, the town experienced a second oil boom as the old wells, then considered marginal, yielded increased production to hydraulic fracturing—the pressurized injection of sand and water into oil strata to remove obstructions and gain a freer flow of petroleum toward the well pumps.[26]

While the productivity of the Corsicana region contributed significantly to future Texas petroleum development, new patterns in the industry eventually brought an end to the area's manufacturing activity. The plant originally built by Cullinan remained the field's sole refinery until 1903. That year, the Central Oil Refining Company (later known as the Richardson-Gay Refining Company) constructed at Corsicana a small refinery with a capacity

26. Matson and Hopkins, *The Corsicana Oil and Gas Field*, p. 247; Warner, *Texas Oil and Gas*, pp. 166–167, 357–359; Taylor, *History of Navarro County*, pp. 180–181.

of less than 100 barrels daily. Some slight diversification in marketing outlets was further obtained by tank-car shipment of crude oil by the Houston Oil Company, purchaser of the Southern Oil properties in 1901, to the Southwestern Oil Company, a small refinery it owned in Houston. In 1904, however, the Corsicana Refining Company, successor to J. S. Cullinan and Company, still dominated local manufacturing operations and handled an estimated 80 percent of the field's production.[27]

But after the Spindletop discovery in 1901, the Beaumont-Port Arthur area of the Texas Gulf Coast became a major refining center with the organization of the Texas, Guffey (later Gulf), Security (later Magnolia), and Sun Oil companies. As production in fields adjacent to this area later declined, these companies, seeking to maintain their favorable access to water transportation, extended crude oil pipelines hundreds of miles to new discoveries in Oklahoma, north Texas, and east Texas, Corsicana was thus given access to the Gulf Coast refining area in 1908, as the Texas Company constructed its Port Arthur-Glen Pool (Oklahoma) crude oil pipeline through the field.[28]

This opened the possibility of a substantial outside market for Corsicana crude oil and started the inevitable decline of the manufacturing process at Corsicana as pipeline systems transported local production to larger, newer, and more efficient refineries in the Texas Gulf Coast region. In 1912, the decline was hastened, as the Magnolia Petroleum Company, which had acquired the refinery built by Cullinan, also linked Corsicana by pipeline with its refinery at Beaumont. As Magnolia expanded into one of the nation's larger, more fully integrated oil companies, the plant at Beaumont, with its access to tidewater, became its major refining center. The Corsicana plant was relegated to a minor role; refining operations gradually slackened until the late 1930s, when they were completely suspended and the installation dismantled.[29]

Today, although Magnolia still owns the refinery site and op-

27. Warner, *Texas Oil and Gas*, p. 90; *Report of the Commissioner of Corporations*, I, 149, 266.

28. Warner, *Texas Oil and Gas*, p. 43.

29. Rister, *Oil! Titan of the Southwest*, p. 92; Warner, *Texas Oil and Gas*, p. 90; "History of the Refining Department of the Magnolia Petroleum Company," p. 11.

erates a pipeline transmission station on it, all that remains of Cullinan's plant at Corsicana is one of the original shell stills, preserved for its historical value, a brick warehouse, and a rusting spur line into the area from the Cotton Belt tracks. Yet there is evidence of permanence, for a slow, nodding pumper works away nearby on one of the old shallow-well leases which brought the oil industry to Texas.

By 1901, a further phase had occurred in the development of the Corsicana oil field. The first phase of exploration and production had been begun by inexperienced and poorly capitalized local operators who were later unable to solve the field's storage and marketing problems. In the second phase, out-of-state capital and managerial direction, at the invitation of local business leaders, brought extensive storage facilities and a solution to the marketing problem through the introduction of the manufacturing or refining process. Local managers, particularly Joe Cullinan, representing this foreign capital, soon acquired further control over raw-material sources and expanded into integrated operations through the organization and consolidation of a large producing company. This process of consolidation—the purchase of existing production from smaller operators—was repeated; additional foreign capital, attracted by the market for crude oil offered by a refining installation, soon acquired another of the field's major production companies. Finally, as a further contribution to field stability, experienced oil men were successful in the initiation of the first state legislation to compel petroleum conservation practices.

Corsicana has been rather briefly dismissed as a mere curtain-raiser to the succeeding Texas petroleum development.[30] If industrial progress can, indeed, be labeled in theatrical terms, it should be emphasized that the Corsicana oil activity was not a brief sketch that preceded an unfolding of the main plot, but a fully-developed "Act One" which set the mood and pace of the ensuing scenario and, more important, introduced the lead players. Joseph S. Cullinan was such a player, now on stage and prepared to assume a further role in Texas petroleum development.

30. Rister, *Oil! Titan of the Southwest*, p. 43.

5. The Shift to Spindletop,
   1901-1902

THE DRAMATIC DISCOVERY of oil in January 1901 at Spindletop
near Beaumont, Texas, two hundred miles southeast of Corsicana,
initiated a phase of flush production which drastically accelerated
the development of the entire American petroleum industry. Despite intensified industrial disorder initially wrought by this production, the efforts of experienced oil men, particularly Joseph
S. Cullinan, would eventually bring a high degree of stabilization
to this development. There was, however, one significant difference between early events at Corsicana and those at Spindletop.
While the finding of oil at Corsicana in 1894 was the accidental
by-product of the quest for water, the Spindletop discovery resulted
from the first attempt in Texas to utilize scientific methods in the
search for petroleum. A resident of Beaumont, Patillo Higgins,
pioneered in applying the rudiments of geology to the study of the
Spindletop area. Later, his observations were expanded and veri-

fied by the work of a European-born mining engineer, Anthony F. Lucas. Joe Cullinan and others would complete the work started by the Spindletop discovery, but that discovery should be rightfully credited to the geologic theories—and dogged perseverance—of these two petroleum pioneers, Higgins and Lucas.[1]

Patillo Higgins was a native of Beaumont, son of a local gunsmith, from whom he apparently inherited a tinkering, inventive curiosity, not only about complex machinery but about the entire universe around him. As a young man, he worked in lumber camps north of Beaumont. In a relatively short time he became a specialist in timberlands, and he later made a comfortable living as a real estate broker and timber consultant. Proposing to launch a brick-manufacturing business in his home town, Higgins traveled north, about 1890, to inspect the latest in plant machinery. He noticed that oil was widely used as fuel in eastern manufacturing establishments and became so interested in this process that he visited the petroleum fields of Pennsylvania and Ohio on the return trip to Texas. Once home, Higgins avidly read all he could find on petroleum geology, corresponded with officials of the United States Geological Survey and the Texas Geological Survey seeking further information on the structure of the Gulf Coast area. He became convinced, although he received little encouragement from these professional sources, that substantial deposits of petroleum would be discovered in the Beaumont area.

Higgins was particularly intrigued by a significant topographical feature four miles south of Beaumont. A circular, mound-like elevation more than a mile in diameter swelled a few easily discernible feet over the broad tidewater prairie. Locals often referred to this elevation, later known as "Spindletop,"[2] as the "Big Hill"

1. Several detailed studies have been made of the early Spindletop development. See Warner, *Texas Oil and Gas*, pp. 19–22, 34–48, 85–88, 187–190; Rister, *Oil! Titan of the Southwest*, pp. 50–72; and James A. Clark and Michel T. Halbouty, *Spindletop* (New York: Random House, 1952), pp. 3–91. Unless otherwise noted, these works are the general sources for the ensuing discussion of Spindletop's development prior to the advent of J. S. Cullinan in the Beaumont area.

2. The name "Spindletop" came into general usage after the discovery of oil and was used to designate the post office established at the oil field. (Clark and Halbouty, *Spindletop*, p. 67.)

or the "Sour Springs Mound," since heavily mineralized water wells and evidences of seeping gas and petroleum had been noted in the area for many years. But Higgins based his belief that the area was rich in petroleum on more than these surface indications. He believed that some type of hard rock thrusting up through weaker strata had caused the surface elevation. He further reasoned that, as these weaker strata were arched by the upthrusting rock, natural beds were formed which trapped deposits of petroleum. Moreover, he estimated that these beds were to be found at a relatively shallow depth: about 1,000 feet. Despite discouragement by professional geologists, Higgins became convinced that Spindletop held the key to major petroleum production. With enthusiasm and persistence, he set about organizing an oil exploration company to confirm his theories.

Perhaps better than he knew, Patillo Higgins, the self-educated layman, was to make a significant contribution to the then relatively rude study of petroleum geology. He had articulated the rudiments of what was to be known later as the salt-dome theory. This theory holds that, during the Permian period of geologic time, when most of the southwestern United States was covered by inland seas, thick salt beds were formed by the eventual evaporation of these waters. These salt beds were in turn overlaid with strata of sedimentary rock. But the salt, of a lower specific gravity than the sedimentary deposits above it, became buoyant and thrust upward, probing for weaknesses in the sedimentary structures. Slowly, inexorably, great shafts or "domes" of salt were formed which rammed their way upward, bulging, tilting, or fracturing the sedimentary shales, sandstones, and limestones. Domes in which the salt core extends almost to the surface are referred to as "piercement" structures and are often evidenced by a slight topographical elevation—such as first caught the notice of Patillo Higgins at Spindletop. At the top and along the steep sides of the salt dome, as it thrust its way through the sedimentary strata, were created anticlimes or entrapped beds of porous rock which accumulated petroleum deposits. Higgins's observations and conclusions were essentially correct, although his fellow-pioneer at Spindletop, Anthony F. Lucas, a mining engineer, is credited with giving the theory professional depth and description. Both of these pioneers, however, were soon to learn that, although their theory was valid, finding the oil posed a problem that would severely tax contemporary methods and equipment. Fortunately for the discoverers of Spindletop, their persistence found initial production at a shallow depth—again as Higgins had predicted: 1,020 feet below the surface, near the "cap rock," or top of the dome. Petroleum deposits entrapped at greater depths in narrow belts along the steep sides of the dome obviously required more advanced geological information and technological procedures to find and render productive. After the "flush," cap-rock production of early Spindletop was exploited, the old dome still yielded valuable "flank" production in 1925 at 4,500 feet and again in 1951 at 9,000 feet.[3]

Higgins's efforts to promote capital to explore the mound area were finally successful in August 1892, as the Gladys City Oil, Gas and Manufacturing Company was organized. This company, with a Beaumont lumberman, George W. Carroll, and an attorney, George W. O'Brien, furnishing most of the capital, acquired 1,077 acres of land for $6,642 in the John A. Veatch League near the western edge of Spindletop. Meanwhile, by using its stock to purchase mineral rights, the company acquired an additional 1,700 acres of leased land adjacent to its fee holdings. For his part in shaping the enterprise, Higgins was appointed general manager, and at first he held a substantial amount of the company's stock. But Higgins, with his characteristic enthusiasm, independently acquired leased acreage near the mound. He soon needed money, and eventually he sold all but two of his original 214 shares to George W. Carroll.

Meanwhile, in February 1893, the Gladys City Company executed an agreement with M. B. Loonie, a Dallas sewage contractor, for a test well to be drilled to a maximum depth of 1,500 feet—unless, of course, as Higgins confidently expected, oil was struck

3. Warner, *Texas Oil and Gas*, pp. 187–188, provides a detailed description of salt-dome formations, although, for the layman, Clark and Halbouty, *Spindletop*, pp. 279–294, with its lucid schematic diagrams is more comprehensible. A concise summary is also found in Max W. Ball, *This Fascinating Oil Business* (New York: Bobbs-Merrill Company, 1940), pp. 59–61.

first. The starting of the well was delayed, however, while Higgins and Loonie quibbled over drilling equipment. Loonie had turned up at the site selected by Higgins with a light rotary machine used for water-well drilling that was, as two authorities colorfully stated, "only slightly bigger than a coffee grinder."[4] Higgins refused to permit a start until heavier equipment was obtained and a major barrier thus appeared which further delayed exploration of the area. Until the Corsicana petroleum development brought advanced drilling equipment to Texas more than three years later, the Spindletop field would refuse to yield her oil treasures.

Loonie finally appeased Higgins by subcontracting the test to Walter B. Sharp, a young driller who had an established reputation for completing deep artesian wells in the Dallas area. Sharp arrived at Spindletop with a heavier rotary rig than Loonie's, won the approval of Higgins, and in March 1893, he began drilling. But even this relatively advanced equipment failed to sink the well deep enough for a conclusive test. The Herculean efforts of Sharp and his crew were of no avail as quicksand slowed drilling progress to inches daily. Drought and water shortages in April and May hampered operation of the rig's hydraulic system. Summer wind and rainstorms blew over the derrick and washed more sand into the well. Finally, in July, with men, equipment, and money growing thin, the drillers stopped, and the well was abandoned at only 418 feet. Sharp took his defeat philosophically, renewed his efforts to improve his drilling equipment, and he obviously learned much from this initial failure. He was shortly to become a successful driller at Corsicana and later, after the 1901 discovery well, he returned to Spindletop to work in close association with Cullinan in further field development.

Nor was the ebullient Higgins crushed by the failure of the Sharp test. While disappointed that a greater depth was not reached, Higgins was heartened and claimed his theories were substantiated by a showing of natural gas encountered only sixty feet beneath the mound surface. He felt that this justified further tests, but the officials of the Gladys City Company and other Beaumont business

---

4. Clark and Halbouty, *Spindletop*, p. 17.

Arnold Schlaet.

John W. Gates.

Corsicana citizens celebrate first shipments of local crude oil, 1896.

J. S. Cullinan's refinery at Corsicana, Texas, under construction, 1898.

Cullinan (far left) and a group of his employees at Corsicana in 1898. Dr. M. P. Cullinan, a brother, is fourth from left. W. C. Proctor and Edwy R. Brown are seventh and eighth from left, respectively. Another brother, Frank Cullinan, is third from right; and standing beside him, fourth from right, is W. T. Cushing. The children are sons of Joseph S. Cullinan, Craig and John.

ew of boom town called Gladys City, adjacent to the Spindletop, Texas,
ld, 1901.

**Eastern part of Corsicana, Texas, in 1901, showing oil activity. Insert: J. S.
Cullinan refinery, Corsicana, Texas, 1901.**

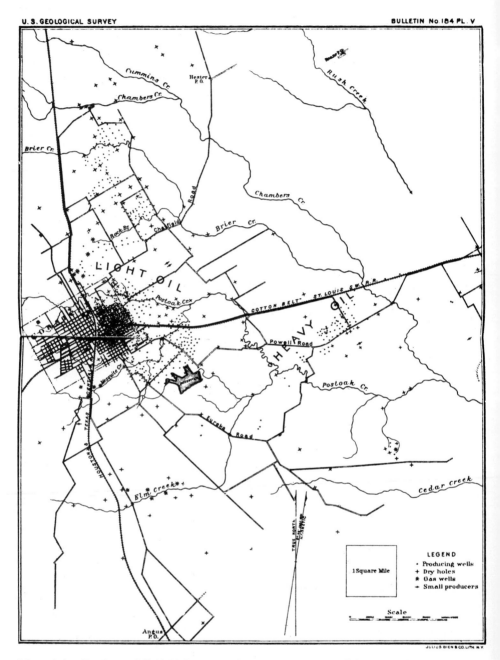

JULIUS BIEN & CO. LITH. N.Y.

Map of the Corsicana Oil Field in 1901. The map was prepared from surveys by the J. S. Cullinan Oil Company.

One of the more famous of the Spindletop gushers, the Heywood No. 2 well, completed in 1901. J. S. Cullinan's Texas Fuel Company bought this well's production at ten cents per barrel.

View of the Spindletop field in 1902, indicating the crowded and indiscriminate drilling practices.

View of fire at Spindletop in 1902. J. S. Cullinan mobilized and directed the fire-fighting effort which saved the entire field from destruction.

Workmen installing a boiler at the Spindletop field, 1902. The boiler was constructed by J. S. Cullinan's Petroleum Iron Works.

Aftermath of Spindletop fire, 1902.

Spindletop field scene, 1902. (Individuals unidentified.)

View of the Sour Lake, Texas, field, circa 1914. The Texas Company tract, purchased and initially developed when J. S. Cullinan served as its president, is in the right background.

Joseph S. Cullinan at age 28. This picture was taken at Lima, Ohio, while Cullinan
was employed by a Standard Oil affiliate.

Joseph S. Cullinan at age 55 (1915).

Joseph S. Cullinan at age 72 (1932).

Walter B. Sharp, president of Producers
Oil Company, the production affiliate of
Cullinan's Texas Company.

James L. Autry at age 54 (1913).

interests, caught up in the economic depression of the mid-1890s, had no capital to support him.

In 1895, however, the Gladys City Company was approached by a West Virginia drilling firm, headed by a team of brothers, Walter A. and James S. Savage. This firm had acquired oil leases at nearby Sour Lake, in Hardin County, and, learning of Sharp's earlier test, they wanted to explore also the Spindletop area. The terms of their proposal seemed particularly attractive to the directors of the Gladys City Company. The Savages would bear all drilling costs and would pay the company a lease bonus plus a one-tenth royalty from any ensuing petroleum income. Higgins opposed this agreement, for it meant that the company would lose the major share of the oil and gas production from its properties. Nevertheless, the directors, with little capital for exploration ventures, saw no other solution and concluded the agreement with the West Virginia firm. Higgins's resentment increased when he saw the type of drilling equipment the Savages were to use in their first test. They brought a percussion or cable-tool rig, which Higgins predicted would be found wholly unsuited to the soft subsurface of the Texas Gulf Coast.

Once again, Patillo Higgins was right. The first test well begun in the fall of 1895 found some gas at shallow depth but the drill then bogged down in quicksand and the well was finally abandoned at 350 feet. When the Gladys City directors persisted in allowing the Savages a second and third test with the cable-tool rig, Higgins disgustedly resigned from the company and sold his few remaining shares of stock. He was right once more: the second and third wells were both abandoned in early 1896 at depths shallower than the 350 feet reached in the first test.

Exploration activity about Beaumont ceased for the next three years, as the Corsicana field monopolized the state's petroleum development. But Patillo Higgins still believed that deeper testing at Big Hill would bring production. His attempts to raise local exploration capital after the previous test-well failures at the mound were unsuccessful, however. Higgins became known as something of an eccentric pest, because of his frequent calls upon Beaumont businessmen urging their financial support. Finally, Higgins decided to seek help from outside sources. Early in 1899, he placed an advertisement in a mining journal, describing the area outside Beaumont and noting its promise for petroleum production. The advertisement brought a reply from Anthony F. Lucas, then living at Washington, D. C. Further correspondence arranged for Lucas to visit Beaumont and inspect the Spindletop area.

The meeting between the ebullient Higgins and the mining engineer brought together two disparate personalities. Anthony Lucas, born in 1855 on a small island along the Dalmatian coast, was the son of Slavic parents named Luchnich. He graduated from a school of mining engineering at Gratz, Austria, later attended the naval academy of the polyglot Hapsburg Empire at Fiume, and was commissioned a lieutenant in the Imperial Navy. In 1879, visiting relatives in the United States, young Luchnich decided to stay, anglicized his name, eventually became a naturalized citizen, and married an American, Caroline Fitzgerald of Macon, Georgia. Lucas had served for many years as a consultant to Louisiana salt and sulphur mining companies, and despite their widely differing backgrounds, Higgins and Lucas soon found themselves in geological agreement as to the oil potential in the area. Lucas, too, had been convinced during his professional studies that the characteristics of Gulf Coast salt domes indicated the presence of petroleum, and more important, he had some capital of his own to expend for an exploratory well at Spindletop.

Through Higgins, Lucas approached the Gladys City Company, which still owned land and held oil leases in the mound area. On June 20, 1899, an agreement was concluded. Lucas agreed to pay the company $31,150 ($11,500 in cash, plus two promissory notes of one and two years' duration) for lease rights to 663 acres. At the same time, Lucas recognized his indebtedness to Higgins by assigning him a one-tenth interest in these leased lands. Within a few weeks, Lucas had hired a Louisiana contractor to begin a test well, using a rotary drilling rig. Again, Higgins warned Lucas that the equipment, suited for coring salt and sulphur at shallow depths, was too light for Spindletop. Nevertheless, Lucas went ahead with the test; and after months of laborious drilling through shifting quicksand, the well was abandoned at 575 feet. But even in this latest failure there was some encouragement. As in the other tests, the drillers not only encountered small pockets of natural gas, but this time they found a few buckets of heavy green crude oil. Lucas wanted to sink another well, heeding Higgins's advice about using heavier equipment, but his funds were depleted by this initial test.

At this point, Lucas traveled east seeking help. He talked with friends in the oil business in western Pennsylvania without success and finally he went on to New York to consult with Henry C. Folger, Jr., of Standard Oil. Folger would not commit himself until he had talked with his partner in the Corsicana enterprise, Calvin N. Payne.

On his next trip to Texas, in February 1900, Payne visited Beaumont with Joseph Cullinan and inspected the Spindletop area. Lucas explained in detail his salt-dome theory, but Payne rather abruptly turned him down, telling him that, despite the previous showings of petroleum, the area would never become profitably productive. Cullinan felt pangs of misgiving over this dogmatic stand taken by Payne. Cullinan had a gambler's instinct, and the thought passed through his mind that the area was at least worth one further test. Yet he did not wish to argue with his friend and benefactor in the presence of Lucas. Moreover, Cullinan's varied Corsicana activities then allowed no capital for personal support of a test well. In short, while the meeting with Payne and Cullinan brought another disappointment for Lucas, its consequences were even more momentous: the country's major petroleum organization, Standard Oil, had refused participation in the exploration of Spindletop.[5]

Within a few weeks, however, the tide turned for Lucas, when he received professional support from Dr. William Battle Phillips, professor of field geology at the University of Texas and director of the state's geological survey. Phillips defied the earlier conclusions of his colleagues, accepted Lucas's salt-dome theory, and sent a letter of recommendation to the Pittsburgh wildcatters, John H. Galey and James M. Guffey, who had initiated the early development of the Corsicana field.

Despite their disenchantment with the pace of the Corsicana development, Galey and Guffey could not resist Phillips's account

5. *Ibid.*, p. 5.

of Spindletop's possibilities. After Galey had made an inspection trip to Beaumont in September 1900, a partnership was organized. Galey and Guffey held the majority share in the enterprise, as they promised to bear the full cost of drilling three test wells. Lucas, needing capital to renew his rights to leased acreage and eager to have his geological theories verified, assigned his oil interests to them for a minority share in the venture. The agreement made no provisions for Patillo Higgins to participate in the partnership, unless Lucas wished to assign him part of his own share. Guffey insisted upon this course, stating that "it would be impractical to cut in everyone who ever heard of oil prospects in Texas."[6]

Backed with capital borrowed from the Mellon banking interests of Pittsburgh, Galey and Guffey wasted little time in completing arrangements for a new test well. A contract was made with the three Hamill brothers, Jim, Al, and Curt, who had all been active in drilling ventures at Corsicana. The Hamills moved their heaviest equipment down to Spindletop for the test and spudded the well on October 27, 1900. They used a twelve-inch rotary rig powered by a thirty-horsepower steam engine but, even then, drilling progress was slow and laborious. In the months that followed, however, the Hamills' crew, with ingenuity backed by brute strength, solved the problems that had defeated earlier tests. They found that iron casing, inserted behind the drilling bit and quickly hammered into place with a 6-by-6-inch beam, prevented loose rock from caving in the well. Curt Hamill got the idea that a heavier-in-density drilling fluid, rather than the clear water used previously at Corsicana, would protect the drilling bit from quicksand and more efficiently flush cuttings to the surface. He then drove a herd of cows through a nearby slush pit to muddy the water and found that the heavier fluid, or "drilling mud," greatly accelerated the well's progress.

Pockets of natural gas and minor traces of oil were encountered before the well reached a depth of 1,160 feet on January 10, 1901.

---

6. *Ibid.*, p. 39. After the discovery of oil at Spindletop, Higgins sued Lucas for $4,000,000 which he claimed was due him for earlier services he had rendered. The suit was settled for a substantially lesser amount, to the satisfaction of all parties, but "Lucas and Higgins . . . had no further relations, either good or bad." (Clark and Halbouty, *Spindletop*, p. 77.)

Drill cuttings from this depth further indicated that the well was nearing the top of the subsurface dome where both Higgins and Lucas had prophesied major petroleum deposits would be found. Suddenly, at 10:30 that morning, with the crew about to lower the drill pipe back into the hole after changing bits, the well broke loose with a roar sending a solid column of oil far above the top of the derrick. Anthony Lucas was soon on the scene, organizing workmen in the construction of earthen dams to confine this cascade of petroleum flowing at a daily rate estimated to be from 75,000 to 100,000 barrels. Before the well was capped nine days later, 800,000 barrels of oil had gushed forth upon the Texas coastal plain dramatically heralding the discovery of the prolific Spindletop field.

News of the Lucas well caught up the Beaumont area, seemingly overnight, in an "oil fever" not witnessed since the petroleum discoveries in western Pennsylvania almost four decades earlier. While Corsicana, with its modest shallow-depth production, had introduced the excitement of petroleum discovery in Texas, Spindletop, with its widely circulated reports of gushers of oil, launched an extended orgy of petroleum speculation. Charles A. Warner notes that four hundred companies with a total authorized capital of over $200,000,000 were organized at Spindletop during the first year of the field's development. By early 1902, oil stock exchanges had been established at Beaumont, Galveston, and Houston to offer the stock of these companies to an eager investing public. Between June 1, 1901, and February 1, 1902, the Houston exchange transacted sales totaling 7,267,570 shares.[7] These companies, of course, ranged from the substantial and legitimate to the blatantly fraudulent, trading upon the gullible investor. Those shorn and disappointed investors would soon give the field another name: "Swindletop."

The major reason for this proliferation of hopes and opportunity at Spindletop was that its original discoverers controlled only a portion of the field's productive acreage. Lucas, Galey, and Guffey held over a thousand acres through lease rights granted by the Gladys City Company, but these claims by no means blanketed

7. Warner, *Texas Oil and Gas*, pp. 45, 188.

the entire mound area. On the northeastern side of the mound a real estate development company, Spindletop Heights, had sold scores of two- and three-acre plots to homeowners. To the southwest, in the John Douthit Survey, there were a number of small farms. In the very center of the mound were further gaps: thirty-three acres held by Patillo Higgins and thirty acres held by the partnership of Keith and Ward. As the proven area of the field was expanded these unclaimed areas became the focal point of spirited, often outrageous, bidding. During the first months of the boom, lease of oil rights went as high as $5,000 for one twentieth of an acre, $15,000 for one twenty-fourth, and $200,000 for two and one-half acres. All of this not only heightened the frenzied pace of financial speculation at Spindletop but also compounded the reckless nature of the field's development. Once again, as at Corsicana, the early Spindletop developers crowded their wells closely together in a mass of wooden derricks that violated the rudiments of field conservation and safety requirements. For example, on one fifteen-acre tract, belonging to the Hogg-Swayne Syndicate, over three hundred wells were drilled.[8]

Under such a high mood of exploration, development proceeded at a rapid pace. By July 1901, there were 14 producing wells in the field with 33 wells under way or about to begin. At the end of December 1901, after almost a year of development, there were 138 productive wells at Spindletop, with 46 additional wells being drilled. At that time, it was estimated that, exclusive of lease and land expenditures, $3,951,085 had been invested in field equipment and installations. Drilling development was even more rapid and frenzied the next year. By October 1902, the field had 440 producing wells. Although the development now encompassed almost the entire mound area, the thickness of derricks on such popular tracts as Hogg-Swayne, Yellowpine, and Keith-Ward resulted in an average spacing of twenty wells to the acre for the field.[9]

Spindletop's crude oil output soon brought Texas into the first rank of the nation's petroleum-producing states. In 1901, the field

8. Clark and Halbouty, *Spindletop*, p. 108.

9. Warner, *Texas Oil and Gas*, p. 40; Clark and Halbouty, *Spindletop*, pp. 112, 119.

produced 3,593,113 barrels of crude oil and ranked fourth in the nation behind the Appalachian field (33,600,000 barrels), the Lima-Indiana field (21,900,000 barrels), and the southern California field (8,800,000 barrels). The next year, Spindletop's production of 17,420,949 barrels moved it to third among the nation's petroleum-producing areas. It trailed the Appalachian field (32,000,000 barrels) and the Lima-Indiana field (23,400,000 barrels), but temporarily, at least, surpassed California's production (14,000,000 barrels).[10]

This flush production launched great changes in the structure of the American petroleum industry, for the Standard Oil combine could not maintain monopolistic control over the nation's crude oil supplies. In 1899, Standard Oil controlled over 85 percent of the annual crude oil output from the nation's two major petroleum-producing areas, the Appalachian and the Lima-Indiana fields. Yet in the Gulf Coast fields, initiated by the Spindletop discovery, that company controlled only 10 percent of the annual crude oil production through 1911.[11]

Several reasons have been offered to explain the reluctance of Standard Oil to compete for major sources of production in the Gulf Coast field. As shown by Calvin Payne's refusal to finance Anthony Lucas in his early Spindletop tests, key Standard executives grossly erred in evaluating the region's petroleum potential. After flush production had been developed, Standard managers still hesitated with doubts as to the commercial value of Gulf Coast crude oil and fears that the unfriendly legal and political climate in Texas might bring antitrust litigation.[12] Furthermore, at that time, Standard Oil was preoccupied with exploration and production activities in the new and promising southern California and Mid-Continental fields.[13] Although Standard Oil was a significant buyer

10. Warner, *Texas Oil and Gas*, p. 375; Harold F. Williamson, Ralph L. Andreano, Arnold R. Daum, and Gilbert C. Klose, *The American Petroleum Industry: The Age of Energy, 1899–1959* (Evanston: Northwestern University Press, 1963), p. 16.

11. Williamson, *et al.*, *The Age of Energy*, p. 7.

12. Hidy and Hidy, *Pioneering in Big Business*, pp. 393–394.

13. New and prolific production also defeated Standard's efforts to monopolize California and Mid-Continental field production. Standard Oil through 1911 controlled only 29 percent of California's annual production and 44 percent of Mid-Continental's annual production. (Williamson *et al.*, *The Age of Energy*, p. 7.)

of crude oil from the beginning of Spindletop development, the company thus lacked permanent control over Gulf Coast production sources. It was not until May 1903 that Standard Oil finally entered the Spindletop area, not by acquiring production properties but through the construction of a small refining installation, the Security Oil Company.[14]

A more important fact was that Standard Oil's absence during the early development of the Gulf Coast fields presented an opportunity for new oil companies to gain a foothold in the nation's industrial structure. Spindletop, with its flush production, was thus to germinate two major petroleum organizations, the Gulf and Texas companies, and revitalize a third, the Sun Oil Company. Within a few years, these firms, and a host of smaller but equally successful companies, would challenge Standard Oil's dominant position within the structure of the American petroleum industry. As Ralph L. Andreano has demonstrated, this inability of Standard Oil to control the production of flush petroleum fields discovered in the early twentieth century effectively destroyed its monopolistic position within the American petroleum industry well before the U. S. Supreme Court ordered dissolution of the organization in 1911.[15]

News of the Lucas well discovery on January 10, 1901, spread like wildfire among the nation's petroleum fraternity. Corsicana oilmen closest to the Beaumont area were the first on the scene. The next day, Joe Cullinan, with a Corsicana oilman, Thomas J. Wood, and a visitor to Texas, Samuel M. ("Golden Rule") Jones, mayor and oilfield equipment manufacturer of Toledo, Ohio, rushed down to inspect the well. The party admired the flowing gusher and Cullinan, who had already telegraphed Standard Oil officials of the discovery, relayed Calvin N. Payne's congratulations. Payne, remembering well his own refusal to support exploration ventures at Spindletop, reportedly asked Cullinan to tell Lucas that "you certainly showed us!"[16]

14. Hidy and Hidy, *Pioneering in Big Business,* p. 393.
15. Ralph L. Andreano, "The Emergence of New Competition in the American Petroleum Industry Before 1911" (Ph.D. dissertation, Economics Department, Northwestern University, 1960), pp. xiii–xiv.
16. Clark and Halbouty, *Spindletop,* p. 62.

During the next weeks, Cullinan was a frequent visitor to the Beaumont area and a keen observer of the progress at Spindletop. He became fully convinced that Standard Oil was committing a serious error in failing to enter the early phases of the field's development. At Cullinan's insistence Standard Oil officials again visited Spindletop in early March 1901. The delegation was led by Daniel O'Day, president of the National Transit Company; also included were Calvin N. Payne and Patrick C. Boyle, then publisher of the *Oil City Derrick* and later of the *Oil and Gas Journal.* Upon conclusion of the inspection trip, O'Day is reported to have remarked: "Too big, too big; more oil here than will supply the world for the next century—not for us!"[17] Whereupon the party departed and, again, Standard Oil officials had refused participation in Spindletop's early development.

Despite Standard Oil's attitude, Cullinan was determined to participate—independently, if need be—in the excitement and opportunity unfolding at Spindletop. But with heavy demands at Corsicana on his time and capital, and the chaotic nature of Spindletop's early development, Cullinan necessarily used a degree of caution before plunging wholeheartedly into new ventures. He decided to organize a small crude oil purchasing company at Beaumont which could be used, if opportunities beckoned, as a base for expanded operations. Meanwhile, Cullinan could retain the income from his Corsicana ventures and await further patterns in the petroleum development at Spindletop.

The charter of Cullinan's new enterprise, the Texas Fuel Company, was filed on March 28, 1901. It was formed to engage in a general oil purchasing business and the company was also empowered to operate a pipeline collecting system. The company's authorized capital was $50,000; the incorporators and directors were listed as Joseph S. Cullinan, his brother, Michael P. Cullinan, and Corsicana oil investor H. L. Scales.[18]

During its first six months, Texas Fuel concluded a few contracts for the purchase of crude oil produced at Spindletop. The com-

17. *Oil and Gas Journal-Oil City Derrick* (Diamond Jubilee Publication, 1934), p. 90.
18. Copy of charter, Texas Fuel Company, Texaco Archives, I, 76–77.

pany's operations were limited and the incorporators made no effort to raise even a substantial portion of the authorized capital. As funds were needed, Cullinan supplied them by drawing cash advances on sales of crude oil from his waterworks lease to the Corsicana Refining Company. During July and August 1901, Cullinan advanced about $1,000 a month to support Texas Fuel's operations.[19]

Texas Fuel, however, enjoyed an initial advantage over the scores of similar companies organized during the early months of the Spindletop field. Through Cullinan, the company had access to a 37,000-barrel-capacity storage tank at the town of Sabine, twenty miles south of the oilfield. This tank, the only substantial storage unit in the entire Beaumont area, had been erected in 1899 by Cullinan's Corsicana enterprises to sell fuel oil from that field to ocean-going vessels entering the Sabine River from the nearby Gulf of Mexico. The tank was erected along the river just north of town on the tracks of the Texas and New Orleans Railroad (Southern Pacific system). Demand for bunkering fuel had never been large and the tank was seldom used until the Spindletop discovery. Then, with the flush production during the first year of the field forcing producers to sell crude oil at three to ten cents a barrel, Cullinan quickly bought up this cheap oil, leased tank cars, and shipped 2,000 barrels daily to this nearby Sabine storage tank to await favorable fuel oil contracts. The Sabine tank was to furnish only a minute portion of the storage needed at Spindletop but it indicated, as at Corsicana, the opportunity offered developers of field collection and storage systems.[20]

Meanwhile, the rapid pace of events indicated that other experienced oilmen sensed opportunity at Spindletop. John H. Galey and James M. Guffey, in front-runners' position because of their agreement with Anthony Lucas reserving the better producing areas, persuaded the Mellon banking interests of Pittsburgh to support them in further Spindletop ventures. In May 1901, they

19. Statement, Corsicana Refining Company to J. S. Cullinan, September 15, 1901, Cullinan Papers.

20. This tank—completely unremarkable in itself—is pictured in William B. Phillips, *Texas Petroleum* (Austin: Mineral Survey, Bulletin No. 1, University of Texas, 1901), p. 4; Burt E. Hull Memo, dated November 10, 1951, Texaco Archives, I, 35.

organized the J. M. Guffey Petroleum Company as a fuel oil marketing firm. The company soon began construction of a field collecting
system and a pipeline to link Spindletop with extensive storage
and loading facilities at Port Arthur, at tidewater only fifteen miles
south. By June 1901, this group had decided to expand operations
by construction of a refinery at Port Arthur. Work on the refinery,
later incorporated as a separate entity and designated the Gulf
Refining Company, began July 13, 1901. By the next spring, work
was completed and the refinery was in operation.[21]

Cullinan watched these developments closely and, it can be
imagined, with a degree of apprehension, for it was obvious that
he could no longer delay a decision on the future role of his Texas
Fuel Company. Further delay, while well-financed and experienced
operators such as the Guffey-Galey-Mellon group began ventures
which might dominate the field's development, threatened the future of his modest venture. Plainly, Cullinan had to find means to
expand quickly the operations and potential of Texas Fuel. Such
an opportunity was the timely offer extended in the fall of 1901
when members of the Hogg-Swayne Syndicate, one of the most
prominent of Spindletop's oil prospecting companies, came to an
agreement with Cullinan concerning the management of their
properties. Through this agreement, Cullinan gained control of
valuable leases at Spindletop, and more important, he won the
further support of the state's most influential business and political
leaders.

The Hogg-Swayne Syndicate was organized in the spring of 1901
as an unincorporated joint venture, or partnership, by the colorful
former Governor of Texas, James S. Hogg, and a group of cronies,
most of them lawyers caught up in the oil fever of Spindletop. The
original members of the syndicate were Hogg, then practicing law
in Austin; James W. Swayne, a Fort Worth attorney who had served
several terms in the state legislature and had been Governor Hogg's
floor leader in the House of Representatives during Texas' twenty-
second legislature (1893–1895); R. E. Brooks, lawyer, of Georgetown,
Texas, who had resigned as a state district judge in 1901 to enter

21. Clark and Halbouty, *Spindletop*, pp. 134–136; Warner, *Texas Oil and Gas*,
p. 46; *Oil Investors' Journal*, May 24, 1902, p.3.

the oil business; A. S. Fisher, also a lawyer from Georgetown, Texas; and William T. Campbell, formerly a newspaper publisher of Lampasas, Texas, who had moved to Beaumont in March 1901, and there organized the Citizens National Bank. Later, the syndicate included Roderick Oliver of Groesbeeck, Texas, a banker who had owned oil leases in the early development of the Corsicana field; E. J. Marshall, a Beaumont attorney; Harris Masterson, a lawyer and landowner of Brazoria County, Texas; and T. E. Moss, Houston attorney and real estate developer.[22]

In July 1901, this group of prominent Texans, determined to share in the oil industry of their state, made a curious agreement with the J. M. Guffey Company. For $180,000, the Hogg-Swayne Syndicate acquired petroleum rights to the J. M. Page fifteen-acre tract located near the southeast edge of the Spindletop mound area.[23] This tract was among the original properties leased by the Guffey-Galey-Lucas partnership and was already surrounded by producing wells. The syndicate had previously paid Page $105,000 for the surface rights to the land that he had paid $450 for in 1897.[24] The Hogg-Swayne group was forced to pay additional amounts to clear title disputes. Eventually they acquired complete or "fee" rights to the tract, but the entire deal cost them $310,000. To pay for this involved transaction, the group had managed to scrape together $40,000 in cash but still owed the balance of $270,000.[25]

Heavily in debt, the syndicate quickly took advantage of the tract's soaring value and during the next few weeks sold off half their purchase in small blocks. Two and one-half acres were sold for $200,000; one-twentieth of an acre, for $50,000; and one twenty-fourth of an acre, for $15,000. Later, as the tract proved highly productive, sales of parcels as small as a thirty-second of an acre were made to purchasers for a small cash payment; the syndicate allowed

22. "The Hogg-Swayne Syndicate," Texas Company Operations Folio, pp. 1–3, Kemp Papers.

23. Assignments of lease, dated July 5, 1901, J. M. Guffey Company to J. W. Swayne, Trustee, Deed Records of Jefferson County, LIII, 140–141. Certified copy in Texaco Archives, I, 92–93.

24. Deed, dated May 23, 1901, J. M. Page to J. W. Swayne, Trustee, Deed Records of Jefferson County, V, 245–246. Certified copy in Texaco Archives, I, 87–88.

25. Clark and Halbouty, Spindletop, p. 108.

the balance to be paid in oil produced. These purchasers, meanwhile, often promoted companies and further divided their parcels for sale.[26]

But the decision of the Hogg-Swayne group to divide and sell brought dire consequences to Spindletop. It resulted in an orgy of frenzied drilling activity with wooden derricks jammed timber-to-timber upon the small lots, referred to as "doormats" by amazed observers. This indiscriminate and wasteful exploitation would in a few months dissipate the mound's reservoir pressure and hasten the end of flush production. In the meantime, this "Onion Patch" of closely packed wells, as the Hogg-Swayne tract was often referred to, quickly assaulted the prolific riches of Spindletop. The result was a glut of production during the summer and fall of 1901 that soon brought extensive storage and marketing problems. The tract's operators, desperate to recoup their high land costs, soon were willing to negotiate contracts to sell crude oil for as little as three cents a barrel.[27]

Thus the decision of the J. M. Guffey Company to assign the J. M. Page tract to the Hogg-Swayne Syndicate was indeed curious, to say the least. Since the Guffey Company held 90 percent of the leases on the hill, it suffered more than any other operator from the ensuing damage to Spindletop's productivity. James Guffey later stated that the sale of the fifteen-acre tract was made because "northern men were not very well respected in Texas in those days. Governor Hogg was a power down there and I wanted him on my side because I was going to spend a lot of money." Guffey further explained that he and the Mellons still feared the possible entry of Standard Oil into the field and welcomed the protection of prominent political leaders such as those included in the membership of the syndicate.[28] Still another explanation offered is that the wise old oil hand, Guffey, knowing full well a chaotic pattern of exploitation was bound to ensue in surrounding holdings, decided to sell part of

26. *Ibid.*, p. 109.

27. *Ibid.*, p. 109–110. However, Clark and Halbouty rationalize that, despite the adverse effects of the Hogg-Swayne sales, the resulting drilling boom "gave hundreds of men the opportunity to learn the oil business from the ground up."

28. Harvey O'Connor, *Mellon's Millions* (New York: John Day Company, 1933), p. 101.

his own lease at a fantastic price while its value was still high.[29]

Nevertheless, capitalizing on the wild financial plunging of Spindletop's early months, the Hogg-Swayne group soon managed to liquidate most of its debts and realized substantial "paper" profits from the exorbitant prices paid for petroleum rights on the Page lease. Exhilarated by their first oil venture, these Texas petroleum novitiates now made plans for more ambitious projects at Spindletop. In the early fall of 1901, the syndicate started to build a collecting and storage system at Spindletop and acquired the right-of-way to construct a pipeline to Port Arthur. There an option was taken for the purchase of a forty-acre tract for a proposed refinery site.[30]

Soon after these ambitious plans had been hatched, the Hogg-Swayne group ran into trouble. The syndicate had sold most of the plots on the Page lease for a nominal cash payment and agreed to accept the remainder of the purchase price in oil. Many of these agreements became virtually worthless in the fall of 1901, when Spindletop's glut of production hopelessly overtaxed the field's storage and marketing facilities. The syndicate was sobered by the threatened loss from the Page lease deals and the further realization that the group had no one within its ranks with extensive experience in the petroleum industry. Several years later, a syndicate member, James W. Swayne, reminisced about this initial naivete. "The members quickly learned there was no money to be made in producing oil," he recalled, "[yet] the Hogg-Swayne Syndicate did not know anything about the oil business." He lamented that the group knew less than "an eight-year-old child that now lives in the Humble field," and that the syndicate finally resolved to "look out for a man that did know the oil business."[31]

The Hogg-Swayne group thus logically soon approached Joseph S. Cullinan with their problems. Cullinan, chaffing over his inactive role in Spindletop's development, welcomed their interest and soon came to an agreement with the syndicate. Through the

29. Clark and Halbouty, *Spindletop,* pp. 108–109.

30. Burt E. Hull Memo, dated November 10, 1951, Texaco Archives, I, 36.

31. *Oil Investors' Journal,* January 5, 1908, p. 19. Swayne refers to the Humble field, located in Harris County, twenty miles north of Houston, which at the time of his statement (1908) was Texas' most active petroleum area.

Texas Fuel Company, he assumed the responsibility of constructing further field collecting and storage facilities and of marketing the production of the Hogg-Swayne properties. The syndicate, through the transfer of the previously acquired pipeline right-of-way and the Port Arthur refinery site option, subscribed to half of the authorized $50,000 capital stock of the Texas Fuel Company. While Cullinan promised to supply most of the additional capital needed for the enterprise, members of the syndicate promised immediately to try to raise funds to start construction of the pipeline and storage projects. The directory of Texas Fuel was expanded to allow for representation of the Hogg-Swayne group. M. P. Cullinan and H. L. Scales were replaced by syndicate members Rod Oliver and R. E. Brooks in December 1901. A month later, E. J. Marshall and William T. Campbell were added to the directory. Officers elected at that time were Cullinan, president; Oliver, vice-president; and Marshall, secretary-treasurer. Contracts were let for the construction of a $27,500 pipeline to Port Arthur and for seven new 37,500-barrel storage tanks costing a total of $68,250 at Spindletop and at Garrison Station, a few miles from the field along the Port Arthur pipeline right-of-way.[32]

Eagerly seizing the opportunity afforded by the Hogg-Swayne agreement, Cullinan quickly expanded the operations of Texas Fuel as a crude oil marketing concern. Company offices, consisting of three rooms rented for $35 a month, were opened in Beaumont, sales agents were employed, and Cullinan began to contract for future delivery of oil by Spindletop producers. At the end of January 1902, Cullinan reported contracts signed for the purchase of 350,000 barrels of crude oil at prices ranging from three to ten cents a barrel. By April 1902, Cullinan had negotiated contracts for future delivery of 1,200,000 barrels of Spindletop crude at an average price of six cents a barrel.[33]

32. W. T. Campbell (Beaumont) to J. S. Cullinan (Beaumont), October 30, 1901, Texaco Archives, I, 116; Hogg-Swayne Agreement, November 9, 1901, Texaco Archives, I, 115; Texas Fuel Company Minutes, Board of Directors' Meetings, Texaco Archives, I, 154–160; pipeline contract with John W. Ennis, December 17, 1901, Texaco Archives, I, 139; tankage contract with Petroleum Iron Works Company, January 4, 1902, Texaco Archives, I, 172–175.

33. Texas Company Operations Folio, pp. 10–15, Kemp Papers; Texas Fuel Company Folder, Autry Papers.

The Hogg-Swayne agreement thus offered Cullinan a wedge for expanded participation in Spindletop's development. As at Corsicana, he entered into the Spindletop field with the blessing of local leaders eager to gain his wide petroleum experience. And again, as at Corsicana, Cullinan vigorously responded in offering collecting and storage facilities and contracts with local producers to take advantage of low-priced flush production. More important for the future development of the Gulf Coast petroleum industry was the fact that the Hogg-Swayne opportunity convinced Cullinan that he should sever his Corsicana connections and concentrate his activities within the Spindletop area. On January 3, 1902, he wrote a letter of resignation from his Corsicana interests. Cullinan was now committed to Spindletop.[34]

But as he expanded his activities at Spindletop, the need for further investment capital became apparent. His plans for a pipeline gathering system, extensive storage facilities, and a refining installation required a substantial outlay of cash—a commodity which Texans of the Hogg-Swayne Syndicate, with most of their capital tied up in Spindletop real estate, certainly could not furnish. Cullinan was thus forced to promote capital from new sources— sources that obviously had to be distinct from his old Standard Oil contacts. That company had refused participation in Spindletop's early development; moreover, the agreement with the prominent Texans of the Hogg-Swayne Syndicate, who had earlier won enhanced political reputations combating the oil trust, provided a further reason why any new capital should be untainted by Standard Oil connections.[35]

The national notoriety of the Spindletop discovery and a wide circle of acquaintances, acquired during a lifetime of association with the petroleum industry, soon brought Cullinan into contact with sources of investment capital. For instance, in the fall of 1901,

34. J. S. Cullinan (Beaumont) to Corsicana Refining Company (Corsicana), January 3, 1902, as noted in "Documentary History of the Texas Company," compiled by Louis W. Kemp and Wilfred B. Talman, Texaco Archives, p. 4.

35. For further information on the role, as attorney-general and governor, of syndicate member James S. Hogg in earlier state antitrust litigation against Standard Oil's marketing affiliate in Texas, the Waters-Pierce Company, see Cotner, *James S. Hogg,* pp. 436–443.

Edward G. Wright, a partner of Cullinan's in the Petroleum Iron Works, Washington, Pennsylvania, made a routine sales call at the office of the Elcho Oil Company at Bradford, Pennsylvania. Elcho, a small producing company, was one of the many interests owned by the wealthy John H. and Lewis H. Lapham, brothers, of New York, who also owned the United States Leather Company, often referred to as the kingpin of the country's "leather trust." The Laphams utilized the crude oil output of Elcho in the tanning process and experimented with it as a bunkering fuel for the vessels of the American-Hawaiian Steamship Company, another Lapham family enterprise. Wright's call at the Elcho office came at the same time as an inspection by Arnold Schlaet, who managed the Laphams' subsidiary investments in oil and shipping. Schlaet listened with growing interest as Wright recounted the latest news from Spindletop gained through recent correspondence with Cullinan. Schlaet was so impressed that he sent an Elcho employee, Charles S. Miller, to Beaumont. Miller knew Cullinan from earlier days when they were both employees of Standard Oil, and he sent Schlaet several favorable reports on the potential of Spindletop and of Cullinan's future plans.

Finally, Schlaet could no longer resist a trip to Spindletop. He arrived at Beaumont in early October 1901, met Cullinan, and observed at first hand the bustling oil activity. He was particularly impressed with Cullinan's foresight in buying up Spindletop's flush production at prices as low as three to five cents per barrel. Because of his close experience with industry and transportation, Schlaet prophesied that there would be an unlimited market for fuel oil, now made even more possible by the Spindletop discovery. He asked the Laphams for capital to help Cullinan build storage and pipeline facilities to take advantage, as he repeated several times with apparent wonderment, "of oil as low as two or three cents per barrel!"[36]

In giving assent to Schlaet's request for capital to back Cullinan

36. Arnold Schlaet (Beaumont) to Lewis H. Lapham (New York), October 9, 1901, Texaco Archives, I, 107–108. This letter also recounts Schlaet's earlier meeting with Wright at the Elcho Oil Company office and his decision to send Charles S. Miller to Beaumont.

and to represent their interests in the Cullinan enterprises, the Laphams were to forge of the two men, Schlaet and Cullinan, a strong managerial team. Nonetheless, it was a team composed of two members with different backgrounds and temperaments. Schlaet, born in Germany in 1859, came to the United States in 1875, and later became a naturalized citizen. He was well educated, spoke excellent English; but he always retained a certain degree of continental aloofness. His experience in the oil business involved solely the supervision of the Laphams' investments. Unlike Cullinan, he had not started at the bottom of the industry as a laborer or boss of a field gang, and he lacked Cullinan's assured touch in working with subordinates, either as a firm disciplinarian or as a jovial comrade recalling earlier experiences in the Pennsylvania fields. Schlaet, however, was to give the team balance, for he exhibited a high degree of Teutonic thoroughness, of methodical perfection, of innate caution often lacking in Cullinan, who was, to a degree, a gambler by instinct.[37]

Schlaet's ability and experience, under Cullinan's direction, thus made a vital contribution to the early success of the Texas Fuel Company and its successor, the Texas Company. Schlaet not only gave Cullinan access to new sources of capital, but later, as an officer of that company, used his wide knowledge of domestic and foreign markets as a sales outlet for the company's crude oil and manufactured products. Their differences in personality, in business policies, eventually would bring conflict; but in these early years, the methodical Schlaet unquestionably played a key role in Cullinan's plans for business expansion at Spindletop.

Shortly after the agreement with the Hogg-Swayne Syndicate gave Cullinan an option to buy a forty-acre refinery site at Port Arthur, Texas, he began negotiations to complete the purchase. The land was owned by the Port Arthur Land and Townsite Company, an enterprise of the colorful and prominent financier, John Warne ("Bet-A-Million") Gates of Chicago and New York. In 1900, Gates acquired control of the Kansas City Southern Railroad, a 778-mile line linking Kansas City with tidewater at Port Arthur. On a visit to that bucolic, rice-growing town soon afterwards,

37. Biographical details from "Employee Biographies" Folio, Texaco Archives.

Gates became interested in its climate and economic potential. He organized the First National Bank, the Port Arthur Rice Milling Company, and the Port Arthur Land and Townsite Company; he expanded the local water and electric companies, and he underwrote municipal bond issues for enlarging the town's school facilities. Gates also viewed the area as a haven from the financial strife of Wall Street. He built a $50,000 summer home at Port Arthur and persuaded his old business crony, Isaac L. ("Uncle Ike") Ellwood to do the same.[38] Clearly, Port Arthur was to take new turns in development with energetic John Gates as a municipal godfather.

In negotiating for the purchase of the refinery site, Cullinan dealt with George M. Craig, Port Arthur manager of the Gates interests. Craig, well aware that the petroleum development at nearby Spindletop was accelerating Port Arthur real estate values, dickered with Cullinan for an increase of the $35,000 purchase price specified in the option agreement. Cullinan was not bluffed. He remarked that a suitable refinery site could be obtained at Sabine Pass, seven miles south of Port Arthur. Craig got the point; the purchase was completed in February 1902, for $35,000.[39]

Impressed with Cullinan's shrewdness, Craig arranged a meeting between Gates and Cullinan during the financier's next trip to Texas. As might be expected, these two men of similar backgrounds and temperaments quickly reached an understanding. Gates, like Cullinan, was largely unschooled; yet he possessed an alert mind, a warm personality, and the flamboyant verve of an inveterate gambler who had risen from work as an obscure barbed-wire drummer to exercise control of the American Steel and Wire Company. Gates was fully informed of the Spindletop development; he had been importuned earlier to invest in several oil ventures there, including the Hogg-Swayne Syndicate.[40] He had resisted these opportunities, however, for Spindletop's chaotic early development, largely in the hands of petroleum amateurs, apparently presented odds too great for even a plunger like Gates. Now, after meeting

38. Lloyd Wendt and Herman Kogan, *Bet A Million! The Story of John W. Gates* (Indianapolis: Bobbs-Merrill Company, 1948), pp. 203–206.

39. George M. Craig (Beaumont) to J. S. Cullinan (Corsicana), February 1, 1902, Cullinan Papers.

40. Cotner, *James S. Hogg*, p. 534.

Cullinan, the odds changed. Gates was impressed with Cullinan's wide experience in the petroleum industry and with the calm, businesslike manner in which he explained his future plans. A bond of trust was quickly but firmly established. Gates, Cullinan reported with satisfaction to friends in Corsicana, soon promised "substantial help for my forthcoming enterprises at Spindletop."[41]

Thus, by early 1902, Cullinan had promises of capital from three different sources: local capital, largely in the form of real estate and oil properties, supplied by prominent Texans of the Hogg-Swayne Syndicate; eastern capital from the Lapham-United States Leather interests of New York and Philadelphia, represented by Arnold Schlaet; and further "foreign" or eastern capital, offered by John W. Gates with his wide holdings and contacts centered in Chicago, New York, and Port Arthur. Shaping and coordinating these sources into a common enterprise required a high degree of tact and diplomacy. Although they shared a common ground of trust in Cullinan's leadership, each of these sources of capital often viewed the others' participation with suspicion and hostility.

The members of the Hogg-Swayne Syndicate had to have guarantees that the use of eastern capital in Cullinan's plans would not overshadow their own interest and place their properties under the control of an out-of-state corporation. Appreciative of the early support they had given the Texas Fuel Company, Cullinan took particular care to reassure the sensitive Texans. He pointed out that, as an adopted son of the Lone Star State, by virtue of his four years' residence there, his interests now coincided with theirs. He promised that any companies chartered in the future would be in full compliance with state law and that the management of such enterprises would be dominated by a directory with a working majority in Texas.[42] Realizing that the members of the Hogg-Swayne group had unquestionably bruised their business egos by the inept pattern of their early Spindletop operations, Cullinan now sought to reinstill confidence. The Texans had learned much from their

41. Cullinan (Beaumont) to W. J. McKie (Corsicana), December 23, 1901, Cullinan Papers.

42. Cullinan (Beaumont) to J. W. Gates (Chicago), March 21, 1902, Texas Fuel Company Folder, Autry Papers.

earlier experiences and Cullinan worked at convincing them that there was no need to shrink from business association with sophisticated Easterners. He confidently predicted that "the Tammany crowd will find their match in the Southerners."[43]

Cullinan soon reassured members of the Hogg-Swayne group and gained their firm support for his plans of business expansion at Spindletop. Ironically, he was to have a much more difficult task achieving co-operation and agreement between the two eastern groups promising capital. Arnold Schlaet, the thorough, conscientious German representing the Lapham interests, took a suspicious—even hostile—attitude concerning John W. Gates's involvement in Cullinan's plans. Whether Schlaet's dislike of Gates sprang from personal dealings or whether Schlaet was prejudiced by stories, many of them apocryphal, of the colorful Gates's alleged business buccaneering and habitual gambling is unknown and relatively unimportant. Schlaet was prepared to believe the worst of stories that pictured Gates as a swindler employing stock manipulation to win control of a vast corporate empire, or of a Gates so confirmed in gambling habits that, bored during a railroad journey, he wagered a million dollars with "Uncle Ike" Ellwood on the speed of raindrops down the coach window.[44]

Schlaet was appalled when he learned that Cullinan was trying to obtain backing from Gates. He wrote Cullinan, in no uncertain terms: "He is not our style, if the newspapers tell the truth about him, and we don't want anybody in who needs watching."[45] When Schlaet found that Cullinan was indeed serious about including Gates in the Spindletop plans, he tried a more direct personal appeal. He was worried about the Gates crowd gaining control and eventually ousting even Cullinan from managerial direction. "Joe," he appealed, on one of the few occasions when he addressed Culli-

43. Cullinan (Beaumont) to W. J. McKie (Corsicana), March 3, 1902, "Documentary History," p. 49.

44. For the traditional, and adverse, view of Gates's career as a "robber baron" see, of course, Matthew Josephson, *The Robber Barons: The Great American Capitalists, 1861–1901*, rev. ed. (New York: Harcourt, Brace & World, Inc., 1962), pp. 372–373, 383, 387, 426.

45. Schlaet (New York) to Cullinan (Beaumont), February 12, 1902, "Documentary History," p. 12.

nan on a first-name basis, "we have got a good thing . . . but you and ourselves [Schlaet and the Laphams] are the only ones . . . who understand this [oil] business and we can run it all the way from end to end without calling for outside help."[46]

Cullinan had previously assured Schlaet that any interest held by Gates in his Spindletop enterprises would be subordinate to the interests held by the Hogg-Swayne and the Schlaet-Lapham groups. To make sure there was no misunderstanding on this point, Cullinan wrote Gates "that there was strong opposition on the part of the local interests . . . to take in other parties, although they are all thoroughly agreeable and anxious to have you personally join on any lines that may be agreeable. . . . " But Gates's interest, Cullinan pointed out, was to be subordinate at all times to those who initiated the enterprise by contributing capital or property "which has grown so steadily in value."[47]

Gates unquestionably understood and was sympathetic toward Cullinan's delicate task in appeasing these diverse capital interests. He relied entirely on Cullinan's direction and was content to relegate his own interest to a subsidiary role if Cullinan's judgment found it necessary. For instance, after the Texas Company had been formed by the union of the three diverse groups of capital, Gates asked Cullinan to use his First National Bank of Port Arthur as the company's major depository and stock registration agent. Cullinan refused; on local banking matters, the company would deal with the Citizens National Bank of Port Arthur, owned by William T. Campbell and Roderick Oliver, both members of the Hogg-Swayne Syndicate.[48] Later, as further evidence of compliance with Cullinan's views, Gates promised to abide by Cullinan's assertion that the directory of the Texas Company was to be dominated by a majority of Texas residents.[49]

46. Schlaet (New York) to Cullinan (Beaumont), February 17, 1902, "Documentary History," p. 12.

47. Cullinan (Beaumont) to Gates (Chicago), January 30, 1902, Texaco Archives, II, 44.

48. Cullinan (Beaumont) to W. T. Campbell (Beaumont), March 3, 1902, "Documentary History," p. 48.

49. Cullinan (Beaumont) to J. W. Gates (Chicago), March 21, 1902, Cullinan Papers.

Since John W. Gates was the most tractable of those groups relying on Cullinan's direction and leadership, his reputation as a reckless gambler and manipulator of corporate affairs was not borne out during his long association with the Texas Company and its predecessor organizations. His correspondence with Cullinan showed him to be a level-headed businessman recognizing Cullinan's ability and willing to follow and support proven business leadership. It was not a blind allegiance, however, for, as a major stockholder, Gates justifiably watched operating details with great interest. He requested monthly reports, noted when they were late, and he queried Cullinan many times for a fuller explanation of items included in company statements. Cullinan, in turn, took care to see that Gates was introduced to the complexities and terminology of the oil business. He even tried to interest Gates's son and sole heir, Charles, in the industry through several long, friendly but apparently unanswered letters.[50] Gates relied on Cullinan's advice on proposed oil investments in New Mexico, Wyoming, and Colorado.[51] The result was a strong bond of personal friendship based on Gates's acceptance and recognition of Cullinan's long experience in the intricacies of the petroleum business.

Cullinan's handling of Gates somewhat mollified Schlaet; the stolid German accepted Gates's participation in the enterprise with a resigned but still wary attitude. "After all," he confided to Elcho's Charles S. Miller, who represented the Lapham interests in Beaumont when Schlaet was needed in the East, "if Cullinan and his immediate friends join with us we would still have a majority . . . against Gates." Still, he fervently hoped "that Cullinan will not be carried away by Gates."[52]

Cullinan further attempted to allay Schlaet's hostility toward Gates by suggesting that the two hold a business conference at Gates's New York office. Schlaet very reluctantly agreed to the meeting. In a spirit of self-sacrifice, he explained to Cullinan that, since

50. Cullinan (Beaumont) to Charles G. Gates (Chicago), April 8, April 30, 1902; May 8, October 24, 1905, Cullinan Papers.

51. J. W. Gates (Chicago) to Cullinan (Beaumont), December 30, 1902; April 23, November 7, December 5, 1903, Cullinan Papers.

52. Schlaet (New York) to Charles S. Miller (Port Arthur), February 10, 1902, "Documentary History," pp. 15, 22.

additional capital was unquestionably needed, "I shall try and see him and be duly impressed by his money and position and hope to make him more anxious to get aboard." Later, he reported that the conference had been satisfactory and that Gates was very eager to become involved in Cullinan's plans.[53] Cullinan had at least won Schlaet's grudging assent to the inclusion of Gates's capital in his Spindletop enterprises.

Three distinct sources of capital were thus combined by Cullinan to finance his plans for expansion at Spindletop. Representing varied business interests, these sources nevertheless united for economic opportunity about the integrity and experience of one key figure, Joseph S. Cullinan. Their complete reliance upon Cullinan's leadership became strikingly clear in September 1902, when a great fire struck the Spindletop field. The fire was started by a driller's carelessly discarded cigar. Flames raced through the crowded wooden derricks and open earthen storage dikes of the Hogg-Swayne "Onion Patch" and threatened to engulf the entire field. If Spindletop was to be saved, someone had to be appointed immediately to marshal workers and direct a co-ordinated fire-fighting effort. Beaumont civic leaders and oil company officials held an emergency meeting and appealed to Cullinan to assume leadership in this crisis. Cullinan agreed, provided he was given authority to enforce his directives at pistol-point if necessary. The power was quickly given. With a week of continuous, backbreaking work, Cullinan and his group of fire-fighters, toiling around the clock, used steam and sand to smother the flames.[54] The field saved, Cullinan, his eyes seared by gas fumes, his body racked with fatigue, finally was forced to bed. Even then, with his eyes bandaged, he held conferences at his bedside with local authorities and dictated a full report of his activities to Schlaet and Gates.[55]

Cullinan recovered his sight within a few days and was soon back to work. Spokesmen for the three groups of capital backing him

53. Schlaet (New York) to Cullinan (Beaumont), February 10, 1902, "Documentary History," pp. 17–18; Schlaet (New York) to Cullinan (Beaumont), March 4, 1902, Cullinan Papers.

54. Clark and Halbouty, *Spindletop*, pp. 117–118.

55. Adrian C. Miglietta (Beaumont) to Schlaet (New York), September 15, 1902, Texaco Archives, VIII, 18.

were naturally relieved that the field was saved but, on second thought, were appalled by his risks of physical injury. John W. Gates urged that he "should use extreme caution for you are the irreplaceable man in our plans."[56] Members of the Hogg-Swayne Syndicate, most of whom were then at Spindletop personally, expressed the same thought.[57] Arnold Schlaet later wrote that "the investment down there . . . will lose every attraction for us should you be disabled."[58] Cullinan's three diverse sources of capital agreed on one point: Joe Cullinan was essential to their Spindletop ventures.

56. J. W. Gates (Chicago) to Cullinan (Beaumont), September 20, 1902, Cullinan Papers.

57. Cullinan (Beaumont) to W. J. McKie (Corsicana), September 23, 1902, Cullinan Papers.

58. As quoted in Marquis James, *The Texaco Story: The First Fifty Years, 1902–1952* (New York: The Texas Company, 1953), p. 18.

6. New Ventures, 1902-1903

WITH ADEQUATE CAPITAL assured, Cullinan quickly moved to expand his Spindletop operations. On January 17, 1902, the charter of the Producers Oil Company was filed and approved by state authorities. This corporation was solely a production company; it was organized for the purpose of "owning, . . . prospecting for, . . . and developing petroleum, oil, and gas . . . " with the added authority to erect and maintain the necessary field equipment for such a business. The incorporators and first directors of the new company were William T. Campbell and Rod Oliver, both of the Hogg-Swayne Syndicate, and George M. Craig, Gates's representative in Port Arthur. Authorized capital was to be $1,500,000 divided into 1,500,000 common shares with a par value of one dollar each, of which 500,000 were to be retained by the company and 1,000,000 were to be available for public subscription.[1]

1. Charter, Producers Oil Company, Texaco Archives, IV, 101.

Although Cullinan was not listed among the incorporators, he was among the original stockholders by subscribing to 100,000 shares.[2] Moreover, he freely acknowledged that he had planned the venture.[3] Recalling the lessons of his Standard Oil apprenticeship and of his recent experience at Corsicana, Cullinan decided to form an oil production and exploration company at Spindletop as soon as possible. His Texas Fuel Company controlled the production of the Hogg-Swayne Syndicate properties, but with the tract's crowded wells, that supply would soon be dissipated. The future of the Texas Fuel Company obviously depended on its access to an adequate supply of crude oil. A production company free to deal with other leaseholders and backed with sufficient funds to explore other areas of the Big Hill seemed the logical solution to Cullinan's problem.[4]

The groups that promised capital for Cullinan's Spindletop plans were immediately willing to go along with this project—that is, all except one, the Lapham group, represented by Arnold Schlaet. Characteristically, the thorough German would not be rushed into a quick decision. From New York, Schlaet asked his Texas representative, Charles S. Miller, to question Cullinan further about the proposed company. It seemed to Schlaet that, with Spindletop's flush production, such a company was superfluous. He felt that Cullinan should concentrate solely on the expansion of Texas Fuel's storage facilities and the execution of long-term crude oil contracts with producers who were forced to sell their output at low prices. As he put it, "I . . . should feel easier if we had our tanks filled and were waiting for tankage rather than oil."[5] He further argued that extension of the Spindletop development and the discovery of new Gulf Coast fields appeared certain. Thus, funds should be retained for rapid extension of collecting pipelines and field storage into these new areas as "the producers will be at the

2. Subscription List, Producers Oil Company, Texaco Archives, IV, 101.
3. Cullinan (Beaumont) to W. J. McKie (Corsicana), January 9, 1902, Cullinan Papers.
4. *Ibid.*
5. Schlaet (New York) to Charles S. Miller (Port Arthur), February 13, 1902, "Documentary History," Supplement, p. 4.

mercy of the transportation companies the same as Spindletop is now."[6]

Cullinan quickly reassured Schlaet that organization of a producing company was both feasible and necessary. Since he had promises of adequate capital for the expansion of the Texas Fuel Company, the incorporation of Producers "by no means indicates that our plans for pipelines and storage will be lessened." Moreover, Cullinan, with his plans for expansion of the Texas Fuel Company still incomplete, thought the quick organization of Producers would retain and whet Gates's interest for future Cullinan ventures. In addition, he pointed out that the organization of a separate corporate entity was necessary under Texas law, which specified that a petroleum-marketing or refining company could not engage in oil production or exploration.[7]

Schlaet was again partially mollified and elected to go along with Cullinan's advice. He soon wrote Charles S. Miller that the Laphams would take a small interest in Producers. This, at least, would allow the suspicious Schlaet to "keep track of matters . . . and to watch that other outside interests do not become of more importance to Cullinan and Campbell than those in which they are interested with us."[8]

On February 10, 1902, Producers Oil held its first stockholders' meeting. Cullinan had in the meanwhile persuaded Gates to assume subscriptions for 52½ percent, or 525,000 shares, of the 1,000,000 shares authorized by the company for public purchase. Gates, in turn, organized a group of his business associates to share this subscription with him. This group included two Chicago bankers, John A. Drake and John F. Harris; and two New Yorkers, James A. Hopkins, president of the Diamond Match Company, and John Lambert, a banker. These men were added to Producers' board of directors, and a motion was passed to assess stockholders an initial 5 percent on each share of stock subscribed. A total initial assessment of $60,000 was collected on 1,200,000 shares by March 1902.

6. Schlaet (New York) to Miller (Port Arthur), February 18, 1902, "Documentary History," Supplement, p. 7.

7. Cullinan (Corsicana) to Schlaet (New York), January 19, 1902, Cullinan Papers.

8. Schlaet (New York) to Miller (Port Arthur), February 11, 1902, "Documentary History," Supplement, p. 3.

Officers elected at the meeting were James Hopkins, president; William T. Campbell, vice-president; Roderick Oliver, treasurer; and George M. Craig, secretary. Although he was not an officer, Cullinan was appointed a member of the company's executive committee. At the same time, the directors approved two agreements: one gave Producers Oil production rights to the Hogg-Swayne tract at Spindletop; the other was an oil contract with Cullinan's Texas Fuel Company. The latter contract specified that Producers was to sell Texas Fuel 1,000,000 barrels of crude oil over a two-year period at twenty-five cents a barrel. With Spindletop crude then selling for three to ten cents a barrel, this contract clearly shows Cullinan's intention not only to control but to spur Producers' future output.[9]

During the summer of 1902, however, the chaotic pattern of Spindletop's development began to take its toll. Wells, particularly in the crowded Hogg-Swayne "Onion Patch," suddenly stopped flowing, as the natural reservoir pressure was dissipated. Walter B. and James S. Sharp, brothers and expert drillers, soon came forth with technical aid to boost production. To overcome the loss of natural pressure, they injected compressed air into the wells to lift the crude oil to the surface. Such artificial methods kept Spindletop's production high throughout 1902 but hastened the end of the field's flush phase. From a high of 17,420,949 barrels of crude oil produced in 1902, Spindletop would fall to 8,600,905 barrels in 1903, to 3,433,842 barrels in 1904, and to 1,652,780 barrels in 1905.[10]

Since Producers relied on the production of the faltering wells in the Hogg-Swayne tract, it found the oil contract with Cullinan's Texas Fuel Company impossible to fulfill. The contract specified that Producers Oil could make up shortages in crude oil deliveries to Texas Fuel by purchases from other sources. But decreased pro-

9. List of "Gates Party" subscriptions, Texaco Archives, IV, 102; list of assessed stockholders, Producers Oil Company, Texaco Archives, IV, 103; Minutes, Board of Directors' Meeting, February 10, 1902, Producers Oil Company, Texaco Archives, I, 207; C. W. Hurd (Beaumont) to Cullinan (Beaumont), March 7, 1902, Cullinan Papers; Prospectus, Producers Oil Company, Texaco Archives, I, 223; Texas Fuel-Producers contract, February 10, 1902 (copy), Texaco Archives, I, 230.

10. Warner, Texas Oil and Gas, p. 375; Clark and Halbouty, Spindletop, p. 153.

duction in the late spring of 1902 temporarily shot crude oil prices to seventy-five cents a barrel.[11] Obviously, Producers, as well as other Spindletop production companies caught in a similar contractual vise, did not relish purchase of seventy-five-cent oil for resale to Texas Fuel at twenty-five cents a barrel.

But Cullinan was prepared to make these companies fulfill their legal obligations—at least, all those companies holding fuel oil contracts with him, except his own Producers. By January 1903, legal counsel for the Texas Fuel Company and its successor, the Texas Company, had filed eighteen suits in the District Court of Jefferson County (Beaumont), Texas, against "non-Cullinan" producing companies at Spindletop which had defaulted on oil contracts at three to ten cents a barrel. However, in the trial of the first of these suits in the summer of 1902, involving a contract between the Independence Oil Company and Cullinan's Texas Company, the court, because of Spindletop's decreasing production, refused to hold the contract valid. This decision, of course, rendered every other long-term oil contract involving such low prices void at the option of the producer.[12]

But Cullinan's absolution of Producers from legal harassment did not bring that company immediate prosperity. It was rapidly slipping into further financial difficulties incurred by the added expense of employing the Sharp brothers' compressed-air process on its Hogg-Swayne wells and the high cost of reserving exploration rights on other parts of Spindletop. The New York and Chicago financiers whom Gates had persuaded to buy Producers stock now refused additional assessments. Cullinan wrote Gates that such action was "premature . . . as the prospects of the company are excellent."[13] Nevertheless, Gates could not prevail upon his eastern friends to continue their support; and by the end of 1902, the company was over $36,000 in debt. At that time, in order to avoid threatened foreclosure, Producers Oil was reorganized. Capitalization was reduced to $50,000, authorized by 10,000 shares of five-

11. Clark and Halbouty, *Spindletop*, p. 153.

12. W. J. McKie (Corsicana) to Cullinan (Beaumont), January 1, 1903, Cullinan Papers; Clark and Halbouty, *Spindletop*, p. 154.

13. Cullinan (Beaumont) to Gates (Chicago), September 19, 1902, Cullinan Papers.

dollar par common stock. For payment of the company's indebtedness, 8,208 shares of stock were transferred to Cullinan, acting as a trustee for the Texas Company, successor corporation to the Texas Fuel Company. The remaining stock, 1,792 shares, was split between the eastern capital backing Cullinan (Lapham-Schlaet-Gates) and the Hogg-Swayne Syndicate members. Walter B. Sharp, the driller who was rapidly making a deserved reputation for initiating new drilling and production techniques at Spindletop, was made president of the company.[14]

Despite the difficulties of Producers Oil during its first year of existence, the firm subsequently justified Joe Cullinan's faith in its potential. It soon became, as originally planned, the exploratory and producing company complementing his major pipeline and refining venture, the Texas Company. Under Walter Sharp, Producers led in developing Sour Lake (in 1903) and other salt-dome structures in the Gulf Coast area, such as the Saratoga, Batson, and Humble fields in Texas and the Jennings and Caddo fields in Louisiana. Between 1902 and 1917, the company produced 73,551,425 barrels of crude petroleum from its Texas-Louisiana holdings and later acquisitions in Oklahoma, Kansas, Wyoming, and New Mexico.[15]

But Cullinan, well aware that the slowdown of Spindletop production would accelerate the search for oil along the Gulf Coast, continued to back other exploration and development companies. One in which he took an interest was the Moonshine Oil Company, organized in May 1902. It was capitalized at $10,000 (one hundred $100-par shares). Cullinan, as trustee for the Texas Company, contributed $5,000; Walter B. Sharp, $1,666.67; James R. Sharp, $1,666.67; and Ed Prather, $1,666.66. Later, Howard R. Hughes, who subsequently invented a bit which revolutionized the industry by making deeper drilling possible, acquired James Sharp's interest. Initially, this company had only one asset: several patent rights to the compressed-air process developed by Walter Sharp to boost Spindletop's faltering production. As other salt-dome

14. Minutes, Board of Directors' Meeting, December 14, 1902, Producers Oil Company, Texaco Archives, IV, 107.
15. Production records, Producers Oil Company, Texaco Archives, IV, 108.

fields, such as Sour Lake and Batson, were developed after Spindletop, Moonshine's process was again widely used. Moonshine's fee was 50 percent of the oil recovered. The company prospered and soon acquired valuable production properties of its own. In 1905, Producers Oil purchased the company for $1,651,763. Moonshine thus rewarded its stockholders with a $16,517.63 liquidating dividend on each share of $100-par-value stock![16]

Another producing company in which Cullinan took an interest was the Paraffine Oil Company. This concern was organized in 1903 by three Beaumont businessmen, William Wiess, Judge William L. Douglass, and S. W. Pipkin. Capitalized at $10,000, the company bought land at Batson Prairie, Texas, twenty miles northwest of Beaumont. In October 1903, the company drilled the discovery well at Batson; and in 1904, Paraffine produced 1,250,000 barrels of crude oil from this field. Since the price of Texas crude was relatively high, as production diminished at older salt-dome fields like Spindletop and Sour Lake, Paraffine made handsome profits the first year of the Batson development. During 1904, the company returned $250,000 in cash dividends to its stockholders on its subscribed capital of $10,000. Later that year, Paraffine acquired production in the Humble, Texas, field and in 1905 pioneered in the development of the North Dayton, Texas, field.[17]

During 1904, Cullinan, both personally and as trustee for the Texas Company, acquired an interest in Paraffine Oil. These combined interests placed 22.2 percent of Paraffine's stock under Cullinan's control. Moreover, Cullinan was given an option to purchase additional stock from one of the company's organizers, William Wiess, should Wiess elect at some future date to dispose of his holdings. This connection between the Texas Company and Parraffine Oil was generally known among the Texas oil fraternity. But in March 1905, Paraffine's directors, fearful that Wiess was seeking to gain control of that company for the Texas Company, forced him to resign.[18]

16. Clark and Halbouty, *Spindletop*, p. 188; Statement of Liquidation, March 17, 1905, Moonshine Oil Company, Texaco Archives, IV, 109.

17. Report to Stockholders, Paraffine Oil Company, August 10, 1905, Cullinan Papers; Larson and Porter, *History of Humble Oil and Refining Company*, p. 27.

18. E. C. Wiess, secretary-treasurer, Paraffine Oil (Beaumont) to Cullinan (Beau-

By that time, however, control of Paraffine was probably immaterial to Cullinan. The company's production was decreasing, and it shared in the steady decline of the Gulf Coast industry during the next few years as petroleum emphasis shifted to new fields in the mid-continental region. In 1909, William Wiess bought out the other stockholders and became president. The company was revitalized through the acquisition of Oklahoma properties. Following the death of William Wiess in 1914, his son, Harry, became president of Paraffine. Young Wiess, later one of the founders of the Humble Oil and Refining Company, used Paraffine properties in the initial capitalization of that firm in 1917.[19]

Still another Cullinan venture into oil exploration and development was the organization of the Landslide Oil Company, formed in 1904, with an authorized capital of $50,000. The first stockholders included former Governor of Texas James S. Hogg, R. E. Brooks, William T. Campbell, and E. R. Spotts of the old Hogg-Swayne group; Walter B. Sharp, of Producers Oil; and Cullinan, whose initial investment was $5,000. The company held lease rights in the then fast-developing Humble, Texas, field. By May 1905, crude oil production from these properties was in excess of 8,000 barrels a day, and at the end of its first year, Landslide had returned 80 percent of its capital in cash dividends to its stockholders.[20]

In these ventures into producing activity, Cullinan was instinctively following a pattern leading to an integrated petroleum operation. It was a pattern that he had learned well during his years with Standard Oil, a pattern he had already initiated successfully at Corsicana. Before his Texas Fuel Company expanded into a substantial pipeline and refinery concern, he realized the operational necessity of insuring for it a constant supply of crude oil. The chaotic nature of the early Gulf Coast development begun by Spindletop in 1901, with its rapid depletion of a once flush field, meant that the head of an integrated operation could not afford

mont), June 15, 1904, Cullinan Papers; William Wiess (Beaumont) to Cullinan (Beaumont), January 23, 1904, Cullinan Papers; *Oil Investors' Journal,* February 1, 1904, p. 7, March 3, 1905, p. 7.

19. Larson and Porter, *History of Humble Oil and Refining Company,* pp. 28, 51.

20. Cullinan (Beaumont) to W. C. Nixon (Galveston), May 31, 1905; W. T. Campbell (Houston) to Cullinan (Beaumont), October 24, 1905, Cullinan Papers.

the slightest apathy toward the activities of exploration and production companies. This explains Cullinan's decision, made with his usual vigor and determination, to organize such a company—Producers Oil—at Spindletop, and later, to develop other fields through close association with the Moonshine, Paraffine, and Landslide ventures.

But Cullinan's participation in these production companies, either through personal stock holdings or as trustee for Texas Fuel and its successor, the Texas Company, caused great concern to his legal advisers. He was often warned that such involvement could be construed as a violation of state incorporation and antitrust laws.

On the face of it, Texas antitrust statutes were particularly stringent; the state had been a pathfinder in this type of regulation. Texas was the second state to pass antitrust laws, in 1889, and it was one of only four states possessing such statutes before the passage in 1890 of federal antitrust legislation in the form of the Sherman Act. The Texas statutes of 1889 defined a trust as a combination of capital, skill, or acts by two or more persons, firms, or corporations for the purposes of restricting trade, limiting production, increasing, reducing, or fixing prices, preventing competition, or refusing to sell or transport. The act included both civil and criminal penalties for violations. Convicted corporations could face outright revocation of charter or fines up to fifty dollars for each day the violating practice continued. Individuals convicted could be fined from $50 to $5,000 and sentenced to imprisonment for one to ten years. In 1899, the statutes were expanded: officers of corporations had to file annual affidavits stating that the firm had not engaged in violations of antitrust statutes, and fines for offending corporations were raised to a maximum of $5,000 for each day the monopolistic practice continued. But the Texas legislature completely rewrote the statutes in 1903, to include periodic amendments passed since the original antitrust law of 1889. Among the significant changes made at that time was the reduction of the maximum corporate fine of $5,000 for each day an abuse continued, imposed in 1899, to the former fine of $50 a day.[21]

21. Tom Finty, Jr., *Anti-Trust Legislation in Texas: An Historical and Analytical*

The ramifications of these stringent antitrust laws for the changing patterns of the Texas petroleum industry were well recognized
by James L. Autry, the very able and conscientious Corsicana
attorney who followed Cullinan to Beaumont as his legal advisor.
Before Spindletop, Autry pointed out that the weight of state antitrust law up to that time had been applied against only one company connected with the petroleum industry, referring to the
prosecution, initiated in 1897, of the Waters-Pierce Oil Company,
Standard Oil's marketing affiliate in Texas. This litigation had
ended in conviction of the Waters-Pierce Company, assessment of
substantial fines, and revocation of the company's charter. But with
the Spindletop discovery and the subsequent extension of petroleum development along the Texas Gulf Coast, Autry warned
that the industry might be subjected to closer examination by state
authorities for possible antitrust violations. He pointed out that
extensive and well-capitalized petroleum concerns such as Cullinan's Texas Company, the Mellons' Gulf companies, and the Pews'
Sun Oil Company, aggressively seeking adequate supplies of crude
oil to complement transportation and manufacturing functions,
"would have to use extreme care to avoid antitrust censure."[22]

Autry later stated that he was particularly concerned with Cullinan's ownership of stock, whether personally or as trustee for the
Texas Company, in several production companies. Cullinan's connections with Producers and Paraffine Oil were particularly well
known in Texas oil circles,[23] and these connections would offer a
strong presumption of a common managerial control designed to re-
strict competition by reserving crude oil supplies for the exclusive
use of Cullinan's marketing and manufacturing ventures. But this
presumption could be overcome, Autry believed, by "dealing with
subsidiary corporations . . . at arms-length." Now, possibly better
aware of the antitrust implications in the advantageous but unexe-
cuted crude oil contract of January 1901 between Cullinan's Texas
Fuel and Producers companies, he cautioned, "I think it important

Review of the Enactment and Administration of the Various Laws Upon the Subject
(Dallas: A. H. Bello & Company, 1916), pp. 1, 7, 11, 13.

22. Autry (Beaumont) to Cullinan (Beaumont), March 29, 1904, Cullinan Papers.
23. *Oil Investors' Journal*, May 24, 1902, p. 1; January 15, 1904, p. 3; May 1, 1904,
p. 2.

that we should not make any contracts with subordinate corporations. . . ."[24]

In addition to possible antitrust difficulties, Autry further
warned that Cullinan's connections with producing companies
might violate Texas incorporation law. The Texas Company, successor corporation to the Texas Fuel Company, had been organized under provisions of the "Pipeline Act of 1899." But Autry
noted that none of the provisions of the pipeline act expressly
authorized a corporation created thereunder to own stock in other
companies. Cullinan's holding of stock in the Producers and Paraffine companies as trustee for the Texas Company was, said Autry,
thus possibly a statutory violation, as "I take it to be a fundamental
proposition that in the absence of statutory authority so to do *one
corporation cannot own stock in another.*"

"Assuming that we have violated the letter . . . of the law,"
Autry wrote that the best defense against future action by state
authorities would be the immediate "sale of our interests in these
properties." But, certainly mindful of the coolness with which
Cullinan would receive such advice, Autry offered a more practical
solution. Since the subsidiary companies "have been deemed almost essential in meeting conditions here . . . and have certainly
served a useful purpose, they should be retained." If, in the future,
state authorities complained, Autry felt that a reasonable time
would be given for Cullinan to eliminate the practice. "If worst
came to worst and a suit was actually instituted," the discontinu-
ance of the practice "would fully answer the state's complaint." As
a precedent, Autry told of former clients charged with antitrust
violations in regard to banking procedures. "I called upon the
Attorney General, made a full show down and confession, practically pleaded guilty for them, as guilty they were, and invoked
his kindliest consideration in giving time for an orderly cessation
of the prohibited practices." The clients were given six months to
comply, this was done, and the "law was vindicated."

This rather pragmatic counsel from Autry matched the *fait
accompli* of Cullinan's plans launched at Spindletop. Through the

24. Autry (Beaumont) to Cullinan (Beaumont), May 28, 1904, Cullinan Papers.
The following quotations from Autry, until otherwise noted, are from this letter.

ensuing years of Cullinan's association with the Texas Company, there was little or no attempt to mask its association with various companies, particularly its major producing subsidiary, Producers Oil. Despite the threat of strictly enforced state incorporation and antitrust laws which might have restricted that relationship, the Texas Company, in a practical sense, was capable of integrated petroleum operations from its inception. It was not reluctant so to inform others. In 1907, the company headed a full-page ad in the *Oil Inventors' Journal* with the boast: "The Texas Company, Producers, Refiners and Distributors; Texas Petroleum and Its Products."[25]

While Cullinan found little practical restriction to his pattern of operations in Texas statutory provisions, he did recognize the threat of strict interpretation. He attempted, particularly in 1905, to influence the Texas legislature to liberalize the incorporation statutes and permit integrated operations under a single corporate charter. Ironically, during his tenure as president of the Texas Company, the legislature would not liberalize these statutes and legitimize Cullinan's practices. In 1917, four years after he left the Texas Company, the state legislature passed the so-called Texas Company Bill, which permitted integrated operations. At that time, Producers Oil was dissolved and its properties absorbed into the mother concern, the Texas Company.[26]

There was one aspect of Texas antitrust law closely followed by state authorities, however. As new and substantial companies gained a foothold in the petroleum industry, following Spindletop, these officials were ever-watchful for a connection between them and America's petroleum "trust," the Standard Oil Company. The political rewards resulting from finding such a connection were obvious; and state officialdom, sharing a suspicion in common with many oilmen, could not believe that Standard would remain aloof from an eventual attempt to control the Gulf Coast petroleum development. Standard Oil's construction, in late 1902, of a small refinery at Beaumont—the Burt Refining Company, later named

25. January 3, 1907, p. 27.
26. Clark, *Chronological History of the Petroleum and Natural Gas Industries*, p. 114.

the Security Oil Company—was well known; but a greater concern was that Standard already secretly controlled Gulf or Texas, both well-capitalized companies obviously bent upon full-scale integrated operations. Thus, both companies were subjected to periodic antitrust prosecution, or threat of prosecution, to reassure state authorities that they were free of Standard Oil taint. In January 1904, state antitrust suits were filed against Gulf, claiming that company was controlled by Standard Oil and that, as a result, it had been a party to agreements designed to destroy competition and to limit production. But further investigation satisfied state authorities that Gulf was not a part of the Standard combine, and the charges were dropped before the case came to trial.[27]

The most publicized attempt to link Cullinan's Spindletop ventures with Standard Oil occurred in 1907, soon after the state had won another antitrust suit against the Waters-Pierce Oil Company and other Standard affiliates within Texas. The state had won an antitrust case against Waters-Pierce in 1897, and that corporation forfeited its right to do business in Texas. In 1900, the company, after reorganization, applied to authorities for a permit to re-enter the state. The permit was granted when officials of the company submitted an affidavit declaring that Waters-Pierce was no longer connected with Standard Oil. Subsequent investigations, however, revealed that Waters-Pierce was still connected with Standard at the time that the re-entry permit was sought in 1900—hence, the trial and conviction again of Waters-Pierce in 1906.

During this trial, it was revealed that United States Senator Joseph W. Bailey had acted as attorney for Waters-Pierce at the time the company sought the re-entry permit in 1900. His political enemies now sprang upon him, sensing a kill, since Bailey was to stand for re-election by the Texas legislature in 1907. A special committee of the state House of Representatives was organized to ascertain whether Bailey's role in the fraudulent concealment of the Waters-Pierce-Standard connection now rendered him unfit

27. *Oil Investors' Journal*, December 15, 1902, p. 1; January 15, 1904, p. 1; February 1, 1904, p. 2.

for re-election. The Bailey investigating committee thus largely became a political forum, generating more heat than light. After several weeks of hearings, the committee finally submitted a report exonerating Bailey from wrongdoing in his past relationships with Waters-Pierce. Meanwhile, the Texas legislature had re-elected Bailey, who had won a Democratic preferential primary the preceding fall, to another term even before the work of the investigating committee was completed.[28]

While the state House committee pursued its investigation, Bailey struck back at his tormentors. In a speech at Sweetwater, Texas, he lashed out at the state attorney general, Robert V. Davidson, as derelict in his duty. Bailey stated that the attorney general should have filed antitrust action against the successor to Cullinan's Texas Fuel Company, the Texas Company, for that company was secretly controlled by Standard Oil. Furthermore, said Bailey, the president of the Texas Company, the former Pennsylvanian, Joe Cullinan, was—an accusation the Senator plainly reserved as the most crushing indictment of all—a Republican.[29]

Attorney Amos L. Beaty, of the Texas Company's legal staff, quickly issued a statement denying that the Texas Company was ever connected with Standard Oil. As evidence of the company's local origins and management, he called attention to the illustrious Texans connected with the Texas Company, including Hogg, Swayne, and Brooks of the Hogg-Swayne Syndicate. Furthermore, asserted Beaty, Texas Company President Cullinan was never a member of the Republican party. He had been a consistent Democrat all his life, even in Pennsylvania, where he served as party chairman in a county "where the majority of his neighbors voted the Republican ticket."[30]

Senator Bailey later revealed the source of his allegation that the Texas Company was controlled by Standard Oil. His information came from a discharged Texas Company employee, A. C. Hall,

28. For further details on Senator Bailey's relationship with Waters-Pierce and the work of the House investigating committee, see King, *History of the Houston Oil Company*, pp. 74–76.

29. *Oil Investors' Journal*, November 19, 1907, p. 7.

30. *Ibid.*

who claimed to have worked with secret records and correspondence indicating that "some one unknown" held a substantial amount of Texas Company stock. James L. Autry, general attorney for the company, promptly invited a full examination of the company's records by state authorities and the accusations were soon forgotten as the Bailey controversy subsided.[31]

Nevertheless, the incident further showed that leaders of the larger companies, such as Texas and Gulf, which germinated at Spindletop, were much concerned about establishing a public image of complete independence from Standard Oil control and welcomed state investigation, if necessary, to establish that image. On the other hand, they counted upon a high degree of public acceptance and official indifference to gain relief from strict interpretation of state incorporation and antitrust laws in the internal management of their firms, particularly in the relationship between producing and transportation or manufacturing subsidiaries.

After the organization of the Producers Oil Company in January 1902, Cullinan turned to the problems of his crude oil marketing concern, the Texas Fuel Company. Capitalized at only $50,000, the company initially suffered from a lack of working capital, since one-half of that amount represented Spindletop lease interests transferred to the company by the Hogg-Swayne Syndicate. But with the promise of substantial financial support from Gates and the Schlaet-Lapham group, Cullinan could plan expansion of the company's field collecting and storage facilities, as well as completion of the ten-mile pipeline from Spindletop to tidewater at Port Arthur. Yet practical considerations tempered Cullinan's ready acceptance of this additional capital to expand his operations through Texas Fuel. The company's charter had to be adjusted to make provisions for increased capitalization and the possible addition of further corporate powers to allow pipeline and manufacturing operations. Meanwhile, with Spindletop in its flush phase demanding constant enlargement of field facilities, time was of the essence. Impatient with corporate technicalities which he felt could be adjusted later, Cullinan prepared to accept the substantial and

31. *Ibid.*, January 5, 1908, pp. 18, 19.

timely capital proffered by Gates and the Laphams and was ready to move ahead with his plans to expand Texas Fuel.[32]

But again Arnold Schlaet counseled caution: he still felt uneasy about Gates's participation in Cullinan's plans. In mid-January 1902, Schlaet pointed out to Cullinan that the unqualified acceptance of Gates's help might give the financier a controlling interest in Texas Fuel and "confront us with the possibility of being put out . . . of our own business." He advised Cullinan to give preference to Lapham and Hogg-Swayne capital and "to keep Gates in line" by conceding him only a minority interest in the Texas Fuel expansion.[33]

Nevertheless, Cullinan reaffirmed that Gates's support was now more essential than ever and that further delay, while other possible sources of capital were contacted, would be ruinous to the Spindletop enterprises. Schlaet, at this point, made his second trip to Texas and spent a week talking with Cullinan and his legal adviser, James Autry. The three agreed on a plan which gave Cullinan the immediate use of the Lapham-Gates capital in the expansion of Texas Fuel. It also insured that he would retain managerial direction while legal arrangements were completed to adjust the company's charter for additional capitalization and further corporate powers.[34]

New investors in the Texas Fuel Company were not to receive shares of stock but "Certificates of Interests." The form and content of these certificates was authorized at a subsequent directors' meeting:

> Received by the Texas Fuel Company (a private corporation) from —————————the sum of——————————Dollars, the same to be used and invested in the business of said company in Texas, and for which the Company acknowledges itself indebted to said——————————.[35]

32. Cullinan (Beaumont) to W. J. McKie (Corsicana), January 15, 1902, Cullinan Papers.

33. Schlaet (New York) to Cullinan (Beaumont), January 18, 20, 1902, Cullinan Papers.

34. Cullinan (Beaumont) to Schlaet (New York), January 21, 1902; Cullinan (Beaumont) to W. J. McKie (Corsicana), January 27, 1902, Cullinan Papers.

35. Minutes, Texas Fuel Company, Directors' Meeting, January 9, 1902, Texaco Archives, I, 190.

Certificate holders were required to accept several conditions. They were to share equally with the stockholders in the profits or losses of the company, but the certificate holders had no voice in the concern's management. Furthermore, the directors reserved the right to retire the certificates and issue stock in the company or in any new companies subsequently organized. The same effect, of course, could have been obtained through issuance of preferred stock or bonds. However, this would have meant complication of Texas Fuel's financial structure at a time when Cullinan and his legal advisors obviously regarded that organization as only a stepping-stone to a more intricate corporate structure.

This form of capital participation was acceptable to all the groups promising financial support. Within the next few months, $450,000 was pledged through the issuance of the certificates, although only about $282,000 in cash was paid in through periodic calls on the certificate holders. Arnold Schlaet, acting for the Laphams, contributed $125,000; the Hogg-Swayne Syndicate, $66,000; Cullinan, $66,000, of which $30,000 was borrowed from John W. Gates; and Gates, himself, $25,000. Cullinan had obviously mollified Schlaet by keeping Gates's participation to a minor share. Gates accepted Cullinan's allocation without complaint, although Schlaet later claimed that Gates wanted an additional $100,000 participation in the venture.[36]

The proceeds from the certificates of interest were quickly used, during the first months of 1902, to expand field collecting and storage facilities. A crew of thirty pipeliners was hired to build a gathering system for the Spindletop leases. This system was then connected to a six-inch pipeline which was constructed to Garrison Station, on the Kansas City Southern Railroad, about two miles east of Spindletop. A ten-acre tract was purchased at that location for a pumping station and tank farm. By late February 1902, crude oil was flowing from Spindletop to Garrison and to storage in six newly constructed 37,500-barrel-capacity tanks. By April 1902,

---

36. Texas Fuel Company Financial Statement, April 1, 1902, Texas Fuel Folder, Autry Papers; Cullinan (Beaumont) to Gates (Chicago), February 10, 1902, Cullinan Papers; Schlaet (New York) to Charles S. Miller (Port Arthur), February 8, 1902, Texaco Archives, II, 160.

nine additional tanks of the same capacity had been constructed. Meanwhile, work was under way on a six-inch pipeline fifteen miles long to link Garrison Station with Port Arthur to the south. Construction of another extensive tank farm was begun at Port Arthur on the forty-acre tract which Cullinan had previously acquired for future use as a refinery site. By April 1902, the company had expended $212,000 in pipeline, storage, and real estate facilities. It had 300,000 barrels of crude oil in storage which cost $14,355, and contracts for future crude oil purchases in excess of 1,100,000 barrels.[37]

Meanwhile, James L. Autry, aided by his Corsicana law partner, W. J. McKie, planned corporate reorganization to complement Cullinan's expanded activities. It was decided to scrap the Texas Fuel Company, with its limited capitalization and corporate scope, and charter a new concern, the Texas Company. But drafting the charter of the new company to fit Cullinan's plans and the requirements of Texas incorporation law again presented practical, as well as legal, problems.

The statutory basis of the new company in Texas corporation law was the so-called Pipeline Act of 1899. This legislation authorized the organization of corporations for the storing, transporting, buying, and selling of oil and gas for heat, light, and other business purposes. Such companies were permitted to make reasonable charges for storage and transportation but were not to discriminate against any producer in the application of these charges and facilities. Further, the companies were granted the right of eminent domain for the condemnation of rights-of-way for pipeline construction.[38]

Since extensive storage and pipeline systems were already under construction in the Spindletop area, this law generally suited Cullinan's business purposes. But complications seemed likely to develop if he went through with his plans to build a refinery at Port Arthur. McKie and Autry pondered whether the sale of refined

37. Burt E. Hull Memo, dated November 10, 1951, Texaco Archives, I, 35–38; Texas Fuel Company Financial Statement, dated April 1, 1902, Texas Fuel Folder, Autry Papers.
38. Tod, *Principal Corporation Laws of Texas,* pp. 8, 9.

petroleum products was permitted under the pipeline law. The close relationship between the Texas Company and such producing concerns as Producers and Moonshine also caused concern, since production was conspicuously omitted in the listing of activities permitted a company chartered under the pipeline law.[39] Moreover, the general incorporation laws of the state prohibited corporations from deviation as to the business purpose defined and described in their charters. At that time—1902—Texas statutes listed sixty different purposes for which a corporation could be organized, and a charter could not contain more than one of these corporate purposes.[40]

Nevertheless, Cullinan's legal counsel prepared and attempted to file with state officials a charter which authorized the Texas Company to engage in both pipeline and producing functions. As the refinery had not been constructed, McKie and Autry felt it premature to open controversy concerning that function. Meanwhile, the attempt to gain permission for both producing and pipeline functions under the same charter was of more immediate importance to Cullinan's plans.[41]

To defend this dual-purpose charter, Cullinan's attorney advanced the opinion that the legislature intended the Pipeline Act of 1899 to be an extension of the Texas statute, Article 642, Section 3A, authorizing the incorporation of petroleum-producing companies. Thus, both pipeline and producing activities could be combined under one corporate charter. McKie conceded to Cullinan that this was a weak argument, since the Texas legislature at the passage of the Pipeline Bill of 1899 had probably intended just the opposite, i.e., to lessen the possibility of unwholesome combination by separating the pipeline and producing functions.[42]

Anticipating rejection of this argument, the attorneys still were not ready to concede defeat. They further pointed out that a Texas

39. McKie (Corsicana) to Cullinan (Beaumont), April 2, 1902, Cullinan Papers.

40. Tod, *Principal Corporation Laws of Texas*, pp. 1–14.

41. McKie (Corsicana) to Cullinan (Beaumont), April 5, 1902, Cullinan Papers.

42. McKie (Corsicana) to John G. Tod, Secretary of State (Austin), April 3, 1902 (copy), Cullinan Papers; McKie (Corsicana) to Cullinan (Beaumont), April 5, 1902, Cullinan Papers.

Supreme Court decision of 1899 permitted the filing of a charter with two distinct corporate purposes in order "to transact one general line of business only."[43] In this case, *Beattie* v. *Hardy*, [44] the primary issue had been the proper designation of a corporation's principal place of doing business. But in an *obiter dictum*, the court directed state officials to accept a charter which allowed a grain storage company also to engage in the buying and selling of agricultural commodities, since such activity was related to the same general business purpose. McKie and Autry now asked that the same reasoning be applied to the charter of the Texas Company:

> We do not contend that we have the right under one corporation to do an oil business, and accumulate and loan money, or build a railroad, because these are separate and distinct lines of business. But we ought to have the right under one large corporation to handle the oil business in all its departments.

The attorneys noted that there was another case then before the Texas Supreme Court which might involve a settlement of this question, but meanwhile it was urged that the charter be accepted "to avoid further delay while awaiting this decision."[45]

Texas Secretary of State John G. Tod agreed, and the charter of the Texas Company, authorizing both production and pipeline functions, was accepted for filing on April 7, 1902. In a subsequent conference with McKie in Austin, Tod indicated that he considered this a conditional acceptance pending the forthcoming Supreme Court decision in *Ramsey et al.* v. *Tod*, the case which Cullinan's attorneys had previously mentioned. Meanwhile, McKie's first contention—that the Pipeline Act of 1899 was a mere extension of Texas law authorizing the incorporation of petroleum production companies—was rejected. McKie thus informed Cullinan that if the original charter of the Texas Company authorizing both production and pipeline activities were to stand, subsequent judicial interpretation would have to uphold the inclusion of both

43. McKie (Corsicana) to Tod (Austin), April 3, 1902, Cullinan Papers.
44. 53 Southwest Reports 685.
45. McKie (Corsicana) to Tod (Austin), April 3, 1902, Cullinan Papers.

these activities within one charter "as absolutely essential to the conduct of any type of petroleum business."[46]

The eagerly awaited decision in *Ramsey et al.* v. *Tod*[47] was announced by the Texas Supreme Court on June 23, 1902. This case involved the charter of a farm equipment sales company. Its organizers had sought to include two business purposes within one charter: the establishment of a general mercantile business which also offered loans to merchandise purchasers. Secretary of State Tod refused to accept the charter for filing on the ground that two different businesses were proposed and Texas statutes specifically limited a charter to one line of business, one corporate purpose. In its decision, the Texas Supreme Court upheld the action of the Secretary of State. The court reviewed the historical development of the state's corporation laws and reiterated the conclusion that Texas statutes did not authorize incorporation for two distinct business purposes.

This decision naturally disappointed Cullinan and his legal advisers, for it obviously threatened to invalidate the Texas Company charter with its provision for both producing and pipeline activities. Characteristically, Cullinan wanted to "challenge the decision . . . since the case does not specifically apply to the oil business."[48] McKie, however, counseled caution. "It would be best," he stated, "to file an amended charter to strike out the portion that authorized the company to produce oil . . ." Furthermore, the question now seemed academic in view of Cullinan's close association with Producers Oil and several other production companies. McKie was apparently ready to let sleeping dogs lie, as he predicted that "the Texas Company will still have ample means to conduct an oil business in all its departments. . . ."[49]

Cullinan's plans to legalize his integrated operations at Spindletop through the charter of the Texas Company were thus altered, although they were not drastically changed. The wording of the charter was changed to comply with statutory provisions which

46. McKie (Corsicana) to Cullinan (Beaumont), May 8, 1902, Cullinan Papers.
47. 69 Southwest Reports 133.
48. Cullinan (Beaumont) to McKie (Corsicana), June 24, 1902, Cullinan Papers.
49. McKie (Corsicana) to Cullinan (Beaumont), June 28, 1902, Cullinan Papers.

legal counsel feared would be strictly enforced and thus possibly subject the company to further scrutiny. But Cullinan, having failed to gain sanction for producing and transportation activities under one charter, merely continued to accomplish the same end through closely associated, but separate, corporate entities.

The new charter, however, did make provision for a substantial increase in capitalization. The authorized capital of the Texas Company was $3,000,000. It consisted of 30,000 $100-par-value shares, of which $1,650,000, or 16,500 shares, was to be issued immediately to stockholders, while the company retained the remaining $1,350,000, or 13,500 shares, as treasury stock. The assets of the old Texas Fuel Company were then revalued at $1,250,000 and sold to the Texas Company for 12,500 shares of that company's stock.[50] The Texas Fuel Company was then liquidated and the new Texas Company stock distributed among both its stockholders and certificate-of-interest holders. Meanwhile, additional cash was raised by the sale of 4,000 shares of Texas Company stock through an underwriting agreement with John W. Gates's Chicago brokerage house, Harris, Gates and Company. The underwriting agreement guaranteed sale of the stock at $40 a share with 5 percent sales commission to the brokerage firm. However, the selling price of this stock ranged from $75 to $95 per share.[51]

The question arises as to the degree of over-capitalization, or "water," in the original stock and assets of the Texas Company. The sale of Texas Fuel's assets to the new company for $1,250,000 clearly shows an optimistic and speculative attitude on the part of Cullinan toward Spindletop's development in the spring of 1902. The assets of Texas Fuel, just prior to re-evaluation and sale to the Texas Company, totaled about $300,000.[52] A realistic evaluation of such intangibles as oil-purchase contracts and lease rights

50. Texas Fuel-Texas Company Contract, dated April 30, 1902, Texas Fuel Folder, Autry Papers.

51. Agreement, Harris, Gates & Company, March 21, 1902, Texas Company Folder, Autry Papers; Charles G. Gates (Chicago) to Cullinan (Beaumont), June 16, 1902, Cullinan Papers.

52. Texas Fuel Company Statement, dated April 1, 1902, Texas Fuel Folder, Autry Papers.

against the background of the Spindletop boom was probably impossible. Yet the re-evaluation of Texas Fuel's properties at $1,250,000, particularly in view of the Texas Company's subsequent earnings in 1902–1903, seems fairly conservative.

Cullinan was certainly more realistic in his evaluation of petroleum properties than were the members of the Hogg-Swayne Syndicate, whose lease rights initially made up the bulk of Texas Fuel's assets. In early February 1902, former Governor James S. Hogg and William T. Campbell sailed for London, hoping to interest British investors in the purchase of the Hogg-Swayne Spindletop properties. They asked $6,000,000 for these holdings, which then consisted of five producing wells.[53] Meanwhile, Cullinan, even with the approaching consummation of his plans for organization of the Texas Company, still wanted his Texas friends satisfied as to the value of their oil holdings and their trip was apparently made with his approval. Hogg and Campbell indicated as much by keeping him informed of their negotiations through frequent cable messages.[54] But after almost three months of fruitless talks with British capitalists, Hogg and Campbell came home in disappointment. Their hopes of selling their Spindletop properties dashed, they were now quite ready to use their properties as a basis for capital participation in Cullinan's Texas Company.[55]

The original ownership of the Texas Company's 16,500 shares of outstanding stock thus reflected the three groups of capital marshaled by Cullinan. One group, that represented by Arnold Schlaet and the Lapham leather interests, held a total of 3,350 shares—2,000 registered to Arnold Schlaet as trustee for the Lapham group; 1,000 owned personally by Lewis H. Lapham; and 350 owned personally by Arnold Schlaet. The other aggregation of Eastern capital, that represented by the interests and contacts of John W. Gates, held 3,140 shares. Of this total, 1,000 shares were held by Gates, 990 by John J. Mitchell of Chicago, 500 by John C. Lambert of Chicago, 250 by John A. Drake of Chicago, 150 by

53. Cotner, *James S. Hogg,* pp. 540–542.

54. Campbell (London) to Cullinan (New York), March 10, 11, 1902; Campbell (London) to Cullinan (Beaumont), April 7, 12, 1902, Cullinan Papers.

55. Cotner, *James S. Hogg,* p. 544.

George B. Harris of Chicago, 125 by John C. Hutchins of Chicago, and 125 by James Hopkins of St. Louis.

But Texas, or local, capital recruited by Cullinan controlled the largest number of outstanding shares, 7,022. Cullinan, personally, owned 2,000 shares and the Hogg-Swayne Syndicate held 1,782 shares. Forty-two Texas residents and old oil friends of Cullinan in Ohio and Pennsylvania held a combined total of 3,240 shares. The largest holders within this group were Walter B. Sharp, 300 shares; James L. Autry, 250; R. E. Brooks, 250; William Wiess, 170; George M. Craig, 125; James A. Edson, 125; and Samuel M. Jones, 125. In addition to the three major groups noted above, sixty-two small-lot holders with non-Texas or "non-oil" connections held a total of 2,988 shares, ranging from seventy-five to two shares.[56]

Assuming the usual support awarded management by these small holders, it was apparent that this initial stock distribution gave Cullinan effective control of the Texas Company. His stock, with that of the Hogg-Swayne Syndicate, other Texas investors, acquaintances within the oil industry, and small investors, totaled 10,010 shares as against 6,490 shares held by the Lapham-Schlaet and Gates interests. Also, Arnold Schlaet's concern that the new company might be dominated from its formation by the "Gates crowd" proved groundless. The combined holdings controlled by Cullinan (10,010 shares) and Schlaet (3,350 shares) totaled 13,360 shares as against 3,140 shares for the Gates group. Cullinan reminded Gates of this as he again pointed out that the new enterprise was to be managed by a board of nine directors, "five of whom have been retained from the Texas Fuel directory in order to have a working majority in Texas."[57]

The original board of directors named in the charter of the Texas Company consisted of Cullinan and four former Texas Fuel Company directors, who were also members of the Hogg-Swayne Syndicate: R. E. Brooks, Rod Oliver, William T. Campbell, and E. J. Marshall. Thus, Cullinan and his Texas associates were a majority against the four remaining directors representing the

56. Stockholders List, dated May 1, 1902, Texas Company Folder, Autry Papers.
57. Cullinan (Beaumont) to Gates (New York), March 21, 1902, Cullinan Papers.

Schlaet-Lapham and Gates groups. Arnold Schlaet and Lewis H. Lapham were named directors. The Gates interests were represented by Gates himself and a close associate, Chicago banker John C. Hutchins.

At the first board of directors' meeting, held May 20, 1902, at Beaumont, officers were elected and the organization of the Texas Company was completed. Cullinan was named president, Arnold Schlaet, first vice president, Rod Oliver, second vice president, and E. J. Marshall, treasurer. Fred W. Freeman, Adrian C. Miglietta, and Thomas J. Donoghue, former employees of the Texas Fuel Company, were named, respectively, secretary, assistant secretary, and assistant treasurer.[58]

But soon after the Texas Company was organized, the expanding nature of the Gulf Coast petroleum industry suddenly brought new challenges to Cullinan and to his company's financial structure. In the fall of 1902, with Spindletop's flush production rapidly diminishing, an eager quest for new production began along the Texas Gulf Coast. Now that the pioneering work of Anthony Lucas and Patillo Higgins had amply demonstrated the productivity of Spindletop's salt dome, oilmen began a search for similar structures and were attracted to the small town of Sour Lake, in Hardin County, twenty-five miles northwest of Beaumont. Just outside Sour Lake, in a wooded area near dark, stagnant ponds reeking of sulphur and fed by heavily mineralized springs, was a slight elevation and other geological signs that indicated a large subsurface dome. The presence of mineral springs and petroleum seepages had given the area a minor reputation since the 1840s as a health resort, or spa. These signs also spurred oil exploration beginning in 1893. During the next six years, minor production was intermittently found at shallow depths and the field at one time supported two primitive crude oil refineries. Production at Sour Lake remained sporadic, however, and the field was forgotten as the Corsicana and Spindletop fields were discovered. But now oilmen were ready to take another look at Sour Lake. There was an obvious demand for crude oil to compensate for the loss

58. Excerpt from minutes, Board of Directors' Meeting, May 20, 1902, Texaco Archives, IV, 39.

of flush production at Spindletop, geologic characteristics were promising, and Sour Lake's closeness to the Beaumont area would facilitate the shift of drilling equipment used and developed at Spindletop. The J. M. Guffey Company, as at Spindletop, now pioneered in the resurgent exploration at Sour Lake. In July 1901, that company had 1,129 acres under lease and had drilled two test wells which were productive at a depth of 900 feet. Early in 1902, the Guffey company linked its Sour Lake leases by pipeline with its field collecting and storage system at Spindletop.[59]

While Guffey was getting a head start at Sour Lake, Cullinan was preoccupied with the formation of the Texas Company. But in January 1903, Cullinan, acting as trustee for the Texas Company, acquired for $20,000, cash, an option from R. E. Brooks and James Roche for the purchase of an 865-acre tract at Sour Lake in the middle of the field's petroleum development. This tract was owned by the Sour Lake Springs Company, and the company's directors knew well its potential: they now asked $1,000,000 for the purchase of full surface and mineral rights ("fee rights") to the tract which they had bought in 1899 for $70,000.[60] Under terms of the purchase option held by Cullinan, three test wells could be drilled on the tract before expiration of the option on April 1, 1903.[61]

Cullinan had already assigned Walter B. Sharp of Producers Oil the task of drilling these test wells on the Sour Lake tract before expiration of the purchase option. The third well, completed in early January 1903, "blew in" as a gusher, far more productive than the Guffey wells, with a rate of flow of from 15,000 to 18,000 barrels of crude oil daily.[62] The completion of this well has inspired an oft-told tale: Sharp, ordered by Cullinan to maintain complete secrecy during the Sour Lake tests, purposefully completed the well on a night so dark and stormy that rain had washed

59. Warner, *Texas Oil and Gas*, pp. 3–4, 22–23, 46, 117, 190.

60. *Ibid.*, p. 24.

61. Assignment of Option, January 20, 1903, E. J. Marshall and James Roche to J. S. Cullinan, Trustee, Cullinan Papers.

62. Charles D. Stillman (Sour Lake) to J. S. Cullinan (Beaumont), January 1, 1903, Cullinan Papers. Stillman, an employee of Producers Oil, was the drilling rig crew boss on this well.

away all traces of the gusher when morning came.[63] Nevertheless, there were reasons why Cullinan could have desired secrecy. News of the tract's potential might have encouraged stockholders of the Sour Lake Springs Company to attempt to void the purchase option. Meanwhile, with Spindletop's production decreasing, the price of oil in that field had inched upward to sixty to seventy cents a barrel late in 1902.[64] Cullinan's Texas Company, with great quantities of oil in storage purchased at from three to ten cents a barrel, would obviously benefit from a stabilized or rising price. News of the prolific Sour Lake test might again, as in the flush days of Spindletop, tumble crude oil prices downward.

But such information could not be withheld from the major stockholders of the Texas Company. Cullinan, undoubtedly seeking to establish further rapport between these adverse groups, immediately informed Schlaet of the Sour Lake discovery and asked him to relay the good news to John W. Gates, then in New York.[65] Later that same day, Charles Gates confirmed that Schlaet had wasted little time in relaying the Sour Lake news. He telegraphed: "Yes, Schlaet has just told me of the gusher."[66]

Satisfying as the Sour Lake discovery was, it still placed a considerable strain upon the resources of Cullinan and the new Texas Company—specifically, the problem of raising $1,000,000, in cash, asked by the owners of the Sour Lake tract. Moreover, the money had to be raised quickly. The directors of the Sour Lake Springs Company, certain to learn eventually of the tract's prolific test well, could hardly be expected to extend the purchase option past the April 1, 1903, expiration date. Indeed, within a few weeks, news of the Sharp test well was well known throughout the oil industry. A trade paper noted the results of the test and the details of the purchase option held by Cullinan. It further reported that the Sour Lake Springs Company "positively refused to renew the option for an additional thirty days" and stated, in an editorial

63. Clark and Halbouty, *Spindletop*, p. 185.

64. Warner, *Texas Oil and Gas*, p. 191.

65. Cullinan (Beaumont) to Schlaet (New York), January 8, 1903 (telegram), Cullinan Papers.

66. Charles G. Gates (New York) to Cullinan (Beaumont), January 8, 1903 (telegram), Cullinan Papers.

aside, that the property at $1,000,000 was "well worth the money asked for it."[67]

Cullinan unquestionably realized at this point that money for the purchase of the Sour Lake tract had to come, in the main, from the Eastern interests backing the Texas Company. Only the substantial resources and contacts of John W. Gates and the Schlaet-Lapham group could raise $1,000,000 in the short period of time left to exercise the option. The Texans of the Hogg-Swayne Syndicate, who had contributed their share of capital to the formation of the Texas Company, largely in the form of Spindletop lease interests, could not be counted on for additional capital, particularly since diminishing production at Spindletop had lessened the value of those property rights.

Thus, plans for financing the Sour Lake purchase were largely a matter to be worked out by Cullinan, Schlaet, and Gates. In a series of hectic conferences held during February and March 1903, at Chicago, New York, and Palm Beach, Florida, the three men finally reached an agreement before the expiration of the option on April 1, 1903. It was agreed that the money would be raised through the sale of the 13,500 $100-par-value shares of stock retained as treasury stock by the Texas Company at its incorporation. Gates, characteristically, was so enthusiastic about Sour Lake prospects that he offered to underwrite the entire issue.[68] This, of course, would not do, as far as Arnold Schlaet was concerned. Schlaet insisted that the Lapham interests keep pace with Gates. The new stock issue was first offered to the shareholders of the Texas Company at eighty dollars a share in lots proportionate to their original holdings. Any unsubscribed stock would then be offered to the stockholders of the Sour Lake Springs Company, who could elect to receive stock in lieu of cash from the proceeds of sale. Finally, the brokerage firm of Harris, Gates and Company agreed to underwrite any remaining unsubscribed stock.[69]

This elaborate planning was scarcely necessary, however. The

67. *National Oil Reporter,* March 28, 1903. Clipping in the Cullinan Papers.

68. Gates (Chicago) to Cullinan (Beaumont), February 21, 1903, Cullinan Papers.

69. Cullinan (Beaumont) to W. J. McKie (Corsicana), February 15, 23, March 17, 24, 1903, Cullinan Papers; Texas Company Folder, Autry Papers.

news of the company's excellent prospects at Sour Lake was enough quickly to sell, or subscribe, the entire issue of 13,500 shares. Of this number, 12,208 shares were subscribed by the original stockholders of the Texas Company; 350 shares were purchased by stockholders of the Sour Lake Springs Company, who chose stock in lieu of cash; and 942 shares were sold through the underwriting agreement with Harris, Gates and Company. But the 12,208 shares purchased by the Texas Company stockholders indicated further penetration of Eastern capital into the company. John W. Gates and his son Charles contributed $325,000 for the purchase of 4,063 additional shares; Arnold Schlaet, as trustee for the Laphams, contributed $350,000 for 4, 375 new shares; Cullinan subscribed $100,-000 for the purchase of 1,250 additional shares; and eighty-five small-lot holders of the Texas Company subscribed to 2,107 shares. But, significantly, William T. Campbell, as trustee for the Hogg-Swayne Syndicate, could scrape up only $33,000 for the purchase of 413 shares.[70]

The money was raised and the 865-acre tract of the Sour Lake Springs Company was conveyed to Cullinan on March 28, 1903, for a total consideration of $900,000. A few weeks before the transfer, Cullinan's attorneys, McKie and Autry, found a flaw in the title held by the Sour Lake Springs Company. A former owner of the land had claimed, in the 1890s, a portion of the tract by adverse possession. McKie and Autry maintained that this claim now meant a clouded title, and, moreover, the attorneys insisted that the officials of the Sour Lake Springs Company had deliberately withheld this information. Legal action was threatened to extend the option until the title could be examined further. The directors of the Sour Lake Springs Company were apparently impressed, or intimidated, by the threatened litigation. They agreed to reduce the cash purchase price to $900,000 in consideration of the Texas Company's posting a $100,000 bond as protection against any further title suit.[71]

70. Stockholders' List, May 1, 1903, Texaco Archives, XIII, 28; Texas Company Folder, Autry Papers.

71. Conveyance, March 28, 1903, Sour Lake Springs Company to J. S. Cullinan,

The purchase of the productive Sour Lake tract insured that the young Texas Company was not to be strangled in its infancy by a lack of crude oil. Cullinan, with obvious relief, wired both Schlaet and Gates from McKie's Corsicana office after the deal was finally completed that the "documents [were] executed and delivered to our entire satisfaction."[72] Both Schlaet and Gates sensed the importance of the occasion in their replies. Gates telegraphed "heartiest congratulations on closing the deal [.] Hope it may prove a second Spindletop."[73] The usual reserve of Arnold Schlaet even broke down. He jubilantly replied: "Telegram received [.] Am glad the big ordeal is over and now a big bottle is in order!"[74]

There was cause for continued celebration as the Texas Company's Sour Lake tract soon became highly productive. The discovery well alone produced 250,000 barrels its first month, and, with crude oil then worth sixty cents a barrel, earned the company $150,000. The Texas Company discoveries soon stimulated frenzied drilling activity on surrounding leases and by September 1903, the Sour Lake field was at its peak, with production of from 50,000 to 60,000 barrels daily. The price of crude oil dropped to fifteen cents a barrel, but by that time the Texas Company had already utilized its earlier production in profitable fuel oil contracts and had constructed extensive tankage to buy this later, cheaper production. The field's flush phase was over by the end of 1903, production declined, and 75 of the 150 wells completed that year had been abandoned. Through a program of selective drilling and proper well spacing, however, the Texas Company tract remained productive long after surrounding leases had been abandoned. The tract was revitalized by deeper drilling in 1916, 1923, and 1935. By the end of 1939, the original Sour Lake tract, purchased by Cullinan

---

Trustee, Texaco Archives, XII, 241; W. J. McKie (Corsicana) to Cullinan (Beaumont), February 25, 28, March 21, 1903, Cullinan Papers.

72. Cullinan (Corsicana) to Schlaet (New York), Gates (Chicago), March 28, 1903 (telegram), Cullinan Papers.

73. Gates (Chicago) to Cullinan (Corsicana), March 28, 1903 (telegram), Cullinan Papers.

74. Schlaet (New York) to Cullinan (Corsicana), March 28, 1903 (telegram), Cullinan Papers.

in 1903, had yielded the Texas Company 27,000,000 barrels of crude oil.[75]

Yet this purchase of the property by expansion of the Texas Company's capitalization raised a disturbing question as to the future role of the Hogg-Swayne Syndicate in the firm's management. Was that group entitled to four of the nine Texas Company directorships, when it had been able to contribute only a modest $33,000 as against the $675,000 in Eastern capital raised by Gates, Schlaet, and the Laphams? Further to complicate the problem, Cullinan's close relationship with the four Hogg-Swayne directors, Roderick Oliver, R. E. Brooks, William T. Campbell, and E. J. Marshall, had been strained because of a dispute arising from a Spindletop contract with the Texas Company.

At the height of the Spindletop development in the summer of 1902, Oliver, Brooks, Campbell, and Marshall acquired, in partnership, independent of their Hogg-Swayne connections, one of Spindletop's many small producers, R. L. Cox and Company, which operated on a "door-mat" lease in the Hogg-Swayne Syndicate tract. This company then executed a contract with Cullinan, agreeing to sell the Texas Company 25,000 barrels of crude oil each month for a period of five months beginning in June 1902. The loss of flush production at Spindletop that summer, particularly on the crowded Hogg-Swayne tract, forced the company to default on its promised deliveries to the Texas Company. Cullinan had previously sued several other Spindletop production companies for similar defaults and the local courts had given the defendants relief by voiding the oil-purchase contracts. Nevertheless, Cullinan directed his attorneys, McKie and Autry, to file suit against R. L. Cox and Company on the additional grounds that its officers had received cash advances and borrowed equipment from the Texas Company which had never been repaid or returned. The pleadings were prepared, but McKie asked Cullinan not to file the suit. The attorney felt there was no doubt of the liability of R. L. Cox and Company, but he suggested it might be wise to let the contract "remain in 'status quo' for if the same people should conclude in the future

75. *Oil Investors' Journal*, January 3, 1906, p. 9; Warner, *Texas Oil and Gas*, pp. 191, 199, 204, 373; company production records, Texaco Archives, XIV, 27.

to try to harm the business interests of the Texas Company the contracts we have on hand will be a dangerous weapon to wield against them." In fact, the contract, as McKie further pointed out, "may be a very good reason for leaving some people off of the board of directors at the next meeting."[76]

Cullinan followed the advice of his attorneys and did not file suit. Even before the Sour Lake purchase, however, he had decided to make changes in the directory of the Texas Company involving the Hogg-Swayne group. He wrote both Gates and Schlaet of his decision. Gates, as usual, deferred to Cullinan by replying, "I leave the entire matter in your hands to do as you think best as you are on the ground and your judgment would be better than mine."[77] Arnold Schlaet briskly concurred by pointing out that originally the "Hogg-Swayne interests had . . . a larger representation than their stock interests warranted." He was emphatic:

If there is any doubt in your mind as to the feeling of some of our Texas friends down there, and that they might not be thoroughly interested, or as much interested, in the success of the Texas Company as in outside ventures, then it should become policy for us to change the board of directors and leave out some of them.[78]

The axe fell on the Hogg-Swayne directors William T. Campbell and E. J. Marshall at the stockholders' meeting of the Texas Company held at Beaumont, November 25, 1902. Campbell was replaced on the Texas Company board of directors by Thomas J. Donoghue, assistant treasurer of the company; Marshall was replaced by Fred W. Freeman, secretary of the Texas Company. Another Hogg-Swayne director, Roderick Oliver, remained until after the Sour Lake purchase; he was replaced by Walter B. Sharp of Producers Oil on November 11, 1903. The remaining Hogg-Swayne representative on the original board, R. E. Brooks, apparently won Cullinan's support for the manner in which he handled, along with James Roche, the Sour Lake tract option. Brooks remained a director until he resigned in May 1904. Later, he was

76. McKie (Corsicana) to Cullinan (Beaumont), October 1, 1902, Cullinan Papers.
77. Gates (Chicago) to Cullinan (Beaumont), October 21, 1902, Cullinan Papers.
78. Schlaet (New York) to Cullinan (Beaumont), October 12, 1902, Cullinan Papers.

re-elected to the board, serving again under Cullinan from November 1907, to January 1913.[79]

Some years later, in an interview, James W. Swayne described Cullinan's action in removing the Hogg-Swayne representatives from the Texas Company board in this general, but incisive, recollection:

> Some members of the syndicate, I think, believed by taking in Cullinan they could use him. They soon found that they were dealing with a master that knew every detail of the oil business and was a master of finance as well. Cullinan very soon snapped the others off the board and was himself in absolute control.[80]

Yet Swayne showed little sign of bitterness: "I regard Mr. Cullinan, then, as now, as one of the finest oil men in the country." Swayne's only regret was that he had been unable to acquire further Texas Company stock during the company's early years, for, as he put it, "today I would be a very rich man." He also noted that some of the original stockholders, including Cullinan, had increased their number of shares by using "certain advantages" denied members of the Hogg-Swayne Syndicate.[81]

Doubtless Swayne was referring to the inability of syndicate members to purchase substantially of the Texas Company stock issued at the time of the Sour Lake tract acquisition. Cullinan had participated in the stock issue through loans from, or underwritten by, Gates,[82] and perhaps members of the Hogg-Swayne group had expected Cullinan to arrange similar financing for them. If this was their hope, Cullinan, who was already doubtful of the business loyalty of certain members of the group, obviously did not help them out. With the Hogg-Swayne capital participation thus becoming subsidiary to other interests, Cullinan moved to reduce further that group's representation on the Texas Com-

79. Excerpts, Stockholders' Meetings, November 25, 1902, November 11, 1903; Directors' Roster, 1902–1913, Texas Company Folder, Autry Papers.

80. Interview, *Oil Investors' Journal*, January 5, 1908, pp. 19–20.

81. *Ibid.*

82. Cullinan borrowed $56,000, at 6 percent interest, from Gates to cover his share of the Texas Company stock issued to buy the Sour Lake property. Gates (Chicago) to Cullinan (Beaumont), February 21, 1903, Cullinan Papers.

pany board of directors. At the same time, he retained control of that board by appointing trusted associates and employees to fill vacancies left by the departing Hogg-Swayne directors.

Thus, in the two years following the discovery of Spindletop in 1901, Cullinan had marshaled the resources to expand his first modest venture in that field, the Texas Fuel Company, into the Texas Company, a substantially capitalized firm contemplating extensive storage, transportation, and marketing operations. While state corporation statutes barred that company from production activities, Cullinan, relying on a lenient application of the law by Texas authorities, openly engaged in integrated operations through control of separately incorporated producing firms such as the Producers Oil and Moonshine Oil companies. With Spindletop's productivity diminishing, Cullinan next raised additional capital from the eastern stockholders of the Texas Company, the Gates and Schlaet-Lapham interests, for the purchase of the prolific Sour Lake tract. Differences between Cullinan and the Hogg-Swayne Syndicate representatives on the Texas Company board of directors and the inability of this group to raise its share of the capital required for the Sour Lake tract purchase prompted Cullinan to replace these directors with close associates and company employees. Despite the heavy preponderance of eastern capital within the Texas Company, that company remained "Texan" in its managerial flavor, but now solely through the domination of Cullinan's leadership.

7. Problems of Leadership, 1903-1911

FROM ITS INCEPTION, the Texas Company, well-capitalized and aggressively led, was a very profitable venture. The prolific Spindletop discovery accelerated national acceptance of petroleum as a fuel and Cullinan's company rode the crest of the wave. Able to obtain marketable supplies of low-cost crude oil through extensive production, transportation, and storage operations, the Texas Company, even during its first year, made substantial profits by selling unrefined crude oil to industrial users eager to gain relief from the higher price of solid fuels.

The company made no fuel oil sales until completion of the Port Arthur pipeline, storage, and dock-loading facilities in July 1902. From then through April 1903 and the close of the Texas Company's first year of operation, the company sold 928,434 barrels of crude oil at a gross price of $604,631, or an average sales price of about sixty-five cents a barrel. But since most of this crude oil had been acquired during Spindletop's flush days, it cost the company only $111,412, or an average cost of about twelve cents a barrel. In addition, the company received $46,646 in income from pipeline and storage charges. After consolidation of all income items and expenses the company had an operating profit of $468,-036 in its first year; and of this amount, $165,000 was subsequently distributed to the stockholders—the Texas Company's first in a series of annual dividends which have continued without interruption since 1903.[1]

The demand for crude oil during the company's first year was particularly stimulated by the development of a new fuel oil market relatively close to the Texas Gulf Coast area. In the fall of 1902, Cullinan received a letter from a New Orleans broker, J. Edward Crusel, pointing out that the sugar plantations along the lower Mississippi River offered a ripe market for low-cost fuel, particularly in their refining operations, where high-priced coke costing from eight to twelve dollars a ton was commonly used.[2] Cullinan immediately authorized Crusel to solicit fuel oil orders and promised him a commission of 5 percent on all initial sales.[3] Crusel subsequently secured fifteen new fuel oil contracts with Louisiana sugar plantations during the last three months of 1902, totaling 173,000 barrels at prices ranging from sixty-three cents to eighty-five cents a barrel. New orders and repeat business during the first four months of 1903 brought total sales to Louisiana plantations to 286,000 barrels by the end of the Texas Company's first year.[4]

In these fuel oil contracts with Louisiana sugar manufacturers, Cullinan took special pains to reassure the customer that a constant supply of crude oil would be periodically delivered, once equipment was converted for the use of liquid fuel. The contract usually specified that should the supplier, the Texas Company, fail to deliver the proper quantities at the times specified, the supplier would reimburse the customer for the purchase of substitute fuels

1. Company sales records, 1902–1903, Texaco Archives, IV, 166–169; Earnings Statement, 1902–1903, Texas Company Folder, Autry Papers.

2. J. Edward Crusel (New Orleans) to Cullinan (Beaumont), October 21, 1902, Cullinan Papers.

3. Cullinan (Beaumont) to Crusel (New Orleans), October 22, 1902, Cullinan Papers.

4. Company sales records, 1902–1903, Texaco Archives, IV, 166–169.

in addition to any amounts that were spent for equipment modifications. The contract also noted, for purposes of assessing any future damage claims, an equivalent between oil and coke fuels. It was usually agreed that $5\frac{1}{4}$ barrels of oil supplied the same amount of heat as one ton of coke.[5]

Stimulated by such favorable terms, these sales of crude oil to Louisiana customers provided a welcome market outlet to the young Texas Company. Yet the problem of transportation costs immediately arose. In a fuel oil purchase contract, it was usually agreed that the seller paid all transportation charges. However, as Cullinan wrote to Schlaet, this was really a moot point. Railroad shipping costs in Texas and Louisiana averaged from thirty-five to fifty cents a barrel, and, regardless of whether seller or purchaser paid the tariff, such charges "obviously dampened acceptance of oil fuel."[6] The solution Cullinan proposed, which was heartily endorsed by Arnold Schlaet,[7] was to make shipments by tanker from Port Arthur to storage facilities at New Orleans and then to distribute oil by barge to sugar refineries along the Mississippi River.[8] In August 1902, the company thus acquired its first property outside Texas by leasing ten acres of land at Amesville, Louisiana, across the river from New Orleans, for terminal and storage facilities. That same month the company paid $4,750 for "Texas Barge No. 1," which operated out of Amesville and, with its 600-barrel capacity, periodically supplied sugar refineries along the river and connecting bayous above New Orleans.[9]

Although Cullinan well realized the advantage offered by the Texas Company's access to water transportation at Port Arthur—

5. See particularly fuel oil contract (copy), dated April 21, 1903, between the American Sugar Refining Company of New Orleans and the Texas Company, Cullinan Papers. This contract was for the purchase of 45,000 barrels at 84 cents a barrel, periodic delivery to be made in quantities of 1,000 barrels between July 1903 and March 1904.

6. Cullinan (Beaumont) to Arnold Schlaet (New York), August 4, 1902, Cullinan Papers.

7. Schlaet (New York) to Cullinan (Beaumont), August 11, 1902, Cullinan Papers.

8. Cullinan (Beaumont) to Schlaet (New York), August 7, 1902, Cullinan Papers.

9. *Report of the Officers and Directors of the Texas Company, November 18, 1902,* Cullinan Papers.

an advantage which, he enthusiastically stated, soon lowered transportation costs to Louisiana to "fractions of pennies"[10]—it was Arnold Schlaet who planned and arranged the details of oil shipments from Port Arthur. Schlaet utilized his long association with the Lapham's American-Hawaiian Steamship Company to gain low charter rates on crude oil shipments from Port Arthur in non-company vessels. Later, as the Texas Company acquired its own vessels, Schlaet's advice as to design and operation was eagerly sought and accepted by Cullinan. Thus, in March 1903, it was Schlaet's responsibility to handle the purchase of the company's first ocean-going tanker. This was the *S. S. Florida*, a converted freighter built in 1887 of 1,900 tons carrying capacity, purchased for $80,000.[11]

Another substantial purchaser of crude oil from the Texas Company during its first year was the Standard Oil Company. The fact that Standard Oil had elected to remain aloof from the Gulf Coast development by no means lessened that company's desire to purchase cheap Spindletop crude for its fuel oil and refining operations.[12] Standard bought heavily from the major Spindletop producers and purchased from the Texas Company alone, in 1903, 356,000 barrels of crude oil.[13]

In these negotiations with Standard, Cullinan took particular pains to avoid any charge of favoritism toward his old employers. Upon executing one sizeable contract with Standard Oil in January 1903, for 256,000 barrels at seventy-one cents a barrel, Cullinan wrote W. J. McKie that he had obtained "top-dollar terms."[14] Examination of sales records supports Cullinan. Contracts executed that month ranged from sixty-three cents a barrel upward to

10. Cullinan (Beaumont) to W. J. McKie (Corsicana), February 7, 1903, Cullinan Papers.

11. Schlaet (New York) to Cullinan (Beaumont), March 21, 1903, Cullinan Papers; *History of the Texas Company's Maritime Department,* prepared by department staff (New York: Texaco Company, 1952), p. 7.

12. During 1903, the average price, at the well, of Appalachian field crude was $1.58; that of Lima-Indiana, $1.16; of Mid-Continent, $1.05; and of Gulf fields, $0.40. (Williamson, *et al., The American Petroleum Industry, 1899–1959,* p. 39.)

13. Company sales records, 1902–1903, Texaco Archives, IV, 166–169.

14. Cullinan (Beaumont) to McKie (Corsicana), January 11, 1903, Cullinan Papers.

a high of eighty-five cents a barrel.[15] Considering the large quantities of crude oil purchased by Standard Oil and its assumption of transportation costs via its own tankers at Port Arthur, Cullinan seems to have driven a fair bargain—at least, from the Texas Company's side—with his old associates.

The Texas Company also found an important market for fuel oil sales among southwestern railroads. Cullinan's early experiments with oil as a locomotive fuel at Corsicana now bore fruit after Spindletop. Texas railroads, particularly those operating along the Gulf Coast, quickly converted to low-cost liquid fuel. In its first year, the Texas Company sold 275,000 barrels of crude oil for railroad fuel purposes. The Southern Pacific was the largest purchaser, buying 220,000 barrels, followed by Gates's Kansas City Southern, 50,000 barrels, with lesser amounts being bought by the Gulf, Colorado and Santa Fe and the Texarkana and Ft. Smith railroads.[16]

These sales in the company's first year were but a harbinger of better things to come. In the company's second year, ending April 30, 1904, sales of fuel oil to railroads totaled over 2,500,000 barrels.[17] With demand for fuel oil firming at an average selling price of $1.02 a barrel, and with sizeable quantities of cheap crude oil in its inventory from the flush Sour Lake tract, the second year of the Texas Company was considerably more profitable than its first. The company's profit totaled $794,250 even after payment of $180,000 in dividends to stockholders.[18]

As this ratio of dividend payments to profits indicates, the directors of the Texas Company in these first years established a policy of retaining a substantial portion of annual income for reinvestment in business expansion. Cullinan emphasized this policy in his report submitted at the first annual meeting of the company's stockholders in November 1902. He stated that "the general policy of

15. Company sales records, 1902–1903, Texaco Archives, IV, 167.

16. *Ibid.*, 166–169.

17. The Southern Pacific system was the largest purchaser, buying 1,550,000 barrels, followed by the Gulf, Colorado & Santa Fe, with 525,000 barrels. (Company sales records, 1903–1904, Texaco Archives, XXXII, 165–166.)

18. Earnings Statement, 1903–1904, Texas Company Folder, Autry Papers.

this company from its inception has been to lay well and permanently the foundation for a permanent pipeline, storage, and manufacturing business."[19]

As proof of this policy, the company had spent $652,000 for field equipment and plant installations by the end of 1902. The twenty-mile pipeline linking the field collecting system at Spindletop with Port Arthur, via pumping stations at Garrison and Nederland, had been completed. Extensive storage facilities had been constructed at Garrison, where there were fifteen 37,500-barrel-capacity tanks; at Nederland, where twelve 37,500-barrel-capacity tanks were located; and at Port Arthur, where seven 37,500-barrel-capacity tanks had been erected and seven additional tanks were in process of construction. To service its Louisiana sugar-refinery accounts, the company had secured a long-term lease on the Amesville, Louisiana, tract used for storage across the river from New Orleans. Two storage tanks had already been erected on this land and work was under way on further terminal and docking facilities. The company had acquired its first vessel, "Texas Barge No. 1," to work out of Amesville and, in addition, had placed orders for the construction of ten railroad tank cars. Work had begun on the refinery at Port Arthur, with the company's directors making an initial appropriation of $170,000 for this installation. The company had long outgrown the three-room office in Beaumont, originally rented by the Texas Fuel Company. In the fall of 1902, the company moved its headquarters to the newly constructed Temperance Hall Building in downtown Beaumont and occupied the entire second floor.[20]

The physical assets of the Texas Company were further expanded in 1903. The Port Arthur pipeline was extended fifteen miles north of Garrison to the Sour Lake field and later that year extended an additional fifteen miles north to the newly developed Batson and Saratoga fields. The company also constructed and operated extensive field collecting and storage facilities in the Jennings, Louisiana, field. The Texas Company's first tanker, the S. S. Flor-

19. *Report of the Officers and Directors of The Texas Company, November 18, 1902,* Cullinan Papers.

20. Financial Statement, dated December 31, 1902, Texas Company Folder, Autry Papers; Burt E. Hull Memo, dated November 10, 1951, Texaco Archives, I, 39–42.

ida, was purchased for $80,000 and two additional oil barges were added to this maritime fleet. Further, the Port Arthur refinery installation was completed in the late fall of 1903. Total assets of the Texas Company at the end of its second year, April 1904, including installations, equipment, oil contracts, and real estate, were valued at $4,464,000.[21]

The beginning of refining operations at Port Arthur, on November 3, 1903, did not greatly change or diversify the Texas Company's sales market during these first years. The high sulphur content and heavy specific gravity of Gulf Coast crude oil made it relatively expensive to manufacture into kerosene and high-grade lubricating oils.[22] Since an immediate market existed for fuel oil, there was little need for the Texas Company to concern itself with this inelasticity. But there was a growing demand, particularly as southwestern railroads converted to oil, for a better grade of fuel oil—one at least fairly free of burner-clogging, heat-enervating impurities. The company met this demand at its Port Arthur plant by manufacturing a refined fuel oil marketed under the name "Solar Oil." The refinery also manufactured large quantities of "gas oil," a liquid distillate used by artificial gas distribution companies to enrich their product. Small amounts of asphalt and heavy lubricating oil, or grease, were manufactured but, until the completion of the Oklahoma pipeline in 1907 brought the company better grade mid-continent crudes, the major product at Port Arthur was refined fuel oil. Nevertheless, the Port Arthur refining installation was extensive. Operations began in 1903 with two six-by-thirty-foot stills, each with a 500-barrel capacity. During 1904, four larger stills, measuring fourteen by one hundred and forty feet, were installed and fourteen additional stills, including four of the newest "steam" type, were constructed in 1905. Refinery

21. Financial Statement, dated April 30, 1904, Texas Company Folder, Autry Papers.
22. The best, or "lightest," of the early Gulf Coast crudes, discovered in the Batson field in 1903, averaged 24° on the Baume specific gravity scale, compared to 36° to 40° for Lima-Indiana field crudes and 43° to 47° for Pennsylvania field crudes. The Sour Lake field produced the "heaviest" of the Gulf Coast crudes, ranging from 14° to 22° Baume but with a sulphur content three times higher than that of the Lima-Indiana field. (Williamson, *et al., The American Petroleum Industry, 1899–1959,* p. 75.)

runs totaled 43,200 barrels in 1903; 318,364 barrels in 1904; 548,136 barrels in 1905; and 813,916 barrels in 1906.[23]

The internal organization and personnel of the Texas Company were soon expanded to meet this increased pace of business activity. By 1904, the organizational structure had evolved into three major departments: the general administrative offices located at Beaumont, which included sections handling accounting, legal, sales, purchasing, railroad traffic, and marine transportation matters; the pipeline department, which included an administrative and engineering staff headquartered at Beaumont with three district offices at the Spindletop, Sour Lake, and Saratoga-Batson fields; and a refining department with its offices located at the Port Arthur works. At the end of 1904, these departments included 208 full-time employees.[24]

Cullinan's wide circle of acquaintances within the country's oil fraternity proved invaluable in recruiting experienced administrative and technical personnel for the Texas Company. There were James L. Autry, Guy Carroll, and Ernest Carroll, who launched careers in the petroleum industry under Cullinan at Corsicana and then followed him on to Spindletop. Autry, as the Texas Company's general attorney, headed the company's legal section and was elected a director in 1905. The brothers Carroll, who came to Beaumont as bookkeepers for Cullinan's Texas Fuel Company, eventually became officers (assistant treasurers) in the Texas Company.[25]

Even more valuable to the recruitment of trained personnel was Cullinan's long association with the Standard Oil Company. Using the lure of high salaries and the opportunity for rapid advancement within a new company, the Texas entrepreneur obtained a number of key executives from the ranks of Standard Oil employees. He persuaded William T. Leman, a Standard Oil refinery manager at Lima, Ohio, to come to Beaumont in 1903 as head of the Texas Company's refining department. Leman, whose annual

23. Port Arthur Refinery Yields, 1903–1910, Texaco Archives, XXII, 75; Louis W. Kemp Memo, dated October 11, 1950, Texaco Archives, I, 2–5.
24. Burt E. Hull Memo, dated November 10, 1951, Texaco Archives, I, 43–48.
25. Employee Biographies Folio, Texaco Archives.

salary at Lima was $4,800, was lured to Texas by a $6,500 annual salary and a personal loan from Cullinan to purchase 200 shares of Texas Company stock.[26]

To assist Leman, Cullinan assigned Ralph C. Holmes, who had joined the company in May 1902. Young Holmes had previously worked for Standard Oil subsidiaries in Olean, New York. He was appointed superintendent of the Port Arthur works in 1903. He later succeeded Leman as manager of the refining department, became a director of the company in 1906, and was made a vice-president in 1913. Cullinan's employment of Holmes was of signal importance to the future of the Texas Company. In 1920, Holmes, in collaboration with a company engineer, Fred Manley, developed and patented a continuous thermal cracking process. The so-called Holmes-Manley process freed the company from reliance upon pre-existing patents and substantially increased the efficiency of its motor fuel production. Holmes eventually became president of the Texas Company in 1926 and chairman of the board of directors in 1933.[27]

Another former Standard Oil employee, William T. Cushing, who had worked briefly for Cullinan at Corsicana, became superintendent of the pipeline department in 1903. Later, in 1907, Martin Moran, another old friend of Cullinan from Buckeye Pipe Line, joined the company as manager of the Oklahoma pipeline division. Thomas J. Donoghue became treasurer of the company in 1903, and later, in 1910, he was made a vice-president and director. Donoghue first met Cullinan while working for Standard Oil in the Titusville, Pennsylvania, oil field. John C. Colligan and Otto Holleran came to the Texas Company in 1903 as foremen in charge of the Sour Lake and Saratoga-Batson pipeline districts, respectively. They were also old friends of Cullinan from his Standard Oil-Pennsylvania days.[28]

Yet Cullinan by no means made the Texas Company a sanctuary for all of his old Standard Oil cronies. He took great pains to satisfy

26. *Ibid.*; Cullinan (Beaumont) to William T. Leman (Pittsburgh), November 29, 1903, Cullinan Papers.

27. Employee Biographies Folio, Texaco Archives.

28. *Ibid.*; Burt E. Hull Memo, dated November 10, 1951, Texaco Archives, I, 43.

himself that the personnel selected "were temperate . . . looked upon their position with the company as permanent . . . and appreciated a Texas way of life."[29] But as the news of the Texas Company's formation spread, Cullinan was deluged with requests for employment from people who had been even remotely associated with him during his earlier career. Today, these letters form an interesting part of the Cullinan Papers and range from discreetly worded inquiries on the crisp letterhead of a Standard Oil official to hand-scrawled, semi literate notes from old Pennsylvania oil field cronies who invariably started their requests with "You remember me, Joe . . . " With a notation in the margin as to the acceptability of the applicant, Cullinan dictated a reply to each letter and, if necessary, tempered refusal by recalling some old oil field incident he had shared with the applicant. In many cases, even though it had not been requested, a check was enclosed by Cullinan, particularly where he knew that age or illness had brought unemployment and financial distress.

Also, it should be pointed out that Cullinan did not always get his man. In 1903, he offered the position of superintendent of the Texas Company's pipeline department to Morris M. McCray, manager of the Lima division of Standard Oil's Buckeye Pipe Line Company. After some weeks of correspondence, Cullinan appeared to have McCray's acceptance. But McCray eventually declined, stating that he had doubts about the Texas climate and "life in the South"; he found that it was impossible "to sever my connections with men whom I have been associated with for a quarter of a century."[30]

Yet Cullinan was well aware that prior experience in the petroleum industry was not an absolute necessity and that the expanding operations of the company could utilize men of diversified business and professional backgrounds. The company's early reliance upon water transportation brought about the appointment in 1903 of a New Orleans shipping agent, William A. Thompson, Jr., to

29. Cullinan (Beaumont) to W. J. McKie (Corsicana), April 21, 1903, Cullinan Papers.
30. Morris M. McCray (Lima, Ohio) to Cullinan (Beaumont), May 28, 1903, Cullinan Papers.

manage the maritime section. Thompson, working closely with Arnold Schlaet in New York, later headed the maritime department. He was elected a director and vice-president and held these offices until his death in 1922.[31]

A young Norwegian-born sailor, Torkild Rieber, joined the Texas Company in July 1905, as captain of an oil tanker. Energetic and personable, "Cap" Rieber was soon transferred to management duties in the marine and refining departments. The employment of Rieber was a further example of Cullinan's ability to select young men capable of assuming future executive responsibilities for the Texas Company. Although Rieber resigned in 1919 to become president of Cullinan's Galena-Signal Oil Company, he rejoined the Texas Company in 1927 to head the foreign operations and marine departments. He became a vice-president in 1928 and served as chairman of the board of directors from 1935 to 1940.[32]

In 1903, Clarence P. Dodge was employed to manage the company's sales and purchasing section. This appointment also indicated that Cullinan recognized that a valuable employee need not come exclusively from an oil background. Dodge was a native Texan. After attending the University of Missouri, he embarked upon a banking career, serving between 1889 and 1901 as cashier of banks in Cameron and Temple, Texas. In 1901, Texas Governor Joseph D. Sayers appointed him as the state's first purchasing agent. Dodge was elected a director of the Texas Company in 1904 and served in that capacity until his death in 1926. He was also named manager of sales for the Oklahoma, Arkansas, and Texas areas in 1905, and secretary of the Texas Company in 1913.[33]

Amos L. Beaty was another native Texan employed by Cullinan who was to achieve a long and successful career with the company. After practicing law for several years at Sherman, Texas, Judge

31. Employee Biographies Folio, Texaco Archives.

32. *Ibid.* Rieber was born in Voss, Norway, in 1882. He attended a nautical academy and, at the age of 17, captained his own ship. A vessel under his command was one of the first tankers to load Spindletop crude oil for shipment from Port Arthur in the fall of 1901. After his resignation from the Texas Company in 1940, Rieber again became associated with the Cullinan-founded American Republics Corporation and later served as president of that organization.

33. Employees Biographies Folio, Texaco Archives.

Beaty joined the Texas Company in 1907 to assist James L. Autry. In 1913, Beaty became general counsel and a director. He became the company's president in 1920 and chairman of the board of directors in 1926.[34]

Thus, during the early years of the Texas Company, Cullinan assembled the nucleus of a diversified managerial group that would guide the company for many decades. Cullinan naturally considered it his right as president of the company he had organized to make all important business decisions after possibly, but not always, consulting Schlaet, Gates, or Autry. However, he encouraged subordinate officers to "assume full responsibility for the routine operations of their departments." This would result, he explained, "in much vexing detail being lifted from the shoulders of the company president" and encourage the "growth of managerial responsibility within the ranks of the officers of the company."[35] Undoubtedly Cullinan's ideas were sound. Many of the junior officers brought into the early Texas Company did win later promotions and provide a continuity of management long after Cullinan had left the firm. He had picked good men and, as he put it, they exhibited a "growth of managerial responsibility."

Yet, even during these first years of the Texas Company, the growth of its managerial structure gradually changed Cullinan's own relationship to the organization. In such a fast-paced business, it was often difficult for subordinate officers to distinguish between "policy" and "routine operations." Moreover, Cullinan was often absent from the Beaumont office for days at a time while he inspected and supervised company operations in other areas of Texas, Oklahoma, or Louisiana. He kept in touch with the Beaumont office during these trips and expected to be consulted when important matters developed. However, his Beaumont associates remember him as rather waspish when a field trip was interrupted by what he considered to be office trivia. It was not that he was slothful toward the detail of paper work which even the company president could not wholly escape. The same office associates tell

34. *Ibid.*
35. Cullinan (Beaumont) to William T. Leman (Beaumont), April 12, 1904, Cullinan Papers.

of Cullinan often coming back from field trips, eating at his desk, and working far into the night over matters that had accrued during his absence.[36] While he had a mind that quickly grasped the complexities of such detail, Cullinan thought of himself not as a professional manager merely coordinating the various functions of a large business structure, but as an oilman-leader whose essential tasks were to direct the finding, storing, transporting, and selling of that commodity. Such an oilman served his organization best when he could personally oversee the whirling rotary, the riveting of tanks, and the laying of field pipelines.

The Texas Company did not suffer, however, from Cullinan's lack of interest in the formalities of organizational structure and internal detail. Responsibility for these functions was gradually assumed by another officer and company vice-president, Arnold Schlaet. When the Texas Company was organized, Schlaet's duties were not clear. That he should be elected an officer and a director, as a representative of the Lapham capital, was unquestioned. It was also agreed that he would maintain a New York sales office. But with his wide business experience gained in association with the Lapham's United States Leather Company, these were hardly duties to keep Schlaet wholly occupied. Yet his presence in the East was required by the Laphams, who still retained his services. Cullinan, however, soon found Schlaet's services valuable even if rendered from the remoteness of New York City. Schlaet's work in obtaining shipping for the young company was welcomed and Cullinan, busy elsewhere and seeking relief from the tedium of office detail, delegated to Schlaet the task of examining and approving the company's growing list of fuel oil contracts. This was work that could be handled through correspondence between New York and Beaumont, and the thorough Schlaet took to his new duties conscientiously. He soon became very well informed about the intricacies and strategies of fuel contract negotiation. For instance, Schlaet advised Cullinan to execute an initial 10,000-barrel fuel oil contract with the St. Louis, San Francisco and Texas Railway at fifty cents a barrel, in order to provide a wedge for future business. His advice was sound, for later that year the railroad purchased

36. Employee Biographies Folio, Texaco Archives.

from the Texas Company an additional 202,000 barrels of fuel oil at an average price of eighty cents a barrel.[37]

Meanwhile, Schlaet had developed enough business for the Texas Company in the East by 1905 to require the construction of waterfront terminal and storage facilities at Bayonne, New Jersey. That same year, Schlaet proposed the company's first foreign sales outlet, and $25,000 was appropriated for the construction of warehouse and storage facilities at Antwerp, Belgium. Furthermore, when the company's marine operations section was organized as a separate department in 1906, its personnel was transferred to New York and placed under Schlaet's supervision.[38]

Thus, in the first years of the Texas Company, Cullinan found Arnold Schlaet helpful in sharing the burdens of administrative detail and the innovation of new policies. Schlaet had progressed from a mere supervisor of detail to a responsible company executive of recognized authority. Cullinan's leadership still dominated the company but the continued expansion of its operations and internal structure would, before long, thrust further pressures upon this type of individual management.

Cullinan was obviously busy throughout these early years, in both field and office, and yet he was forced to devote considerable time to another problem of vital importance to the immediate future of the Texas Company: the possibility of merging that company with the Guffey-Gulf Companies. These companies, it will be recalled, were also spawned at Spindletop. The J. M. Guffey Petroleum Company, a producing and crude oil marketing concern, was incorporated in May 1901, to take over the properties of the J. M. Guffey Company, the partnership of John Galey, James M. Guffey, and Anthony F. Lucas, which had drilled the discovery well at Spindletop. Meanwhile, Guffey used additional funds advanced by the Mellon banking house of Pittsburgh, to buy the interests held by his erstwhile partners, Galey and Lucas. Thus, the major stockholders of the J. M. Guffey Petroleum Company were Guffey, who was its president and operational head, and the Mellon family,

37. Schlaet (Washington, D.C.) to Cullinan (Beaumont), March 29, 1904, Cullinan Papers; company sales records, 1903–1904, Texaco Archives, XXXII, 166.

38. Burt E. Hull Memo, dated November 10, 1951, Texaco Archives, I, 42; *History of the Texas Company's Marine Department*, pp. 10–11.

whose members Andrew W. and Richard B. held directorships in the company. In November 1901, this group organized a separate corporate entity, the Gulf Refining Company, to operate a refinery at Port Arthur. Guffey was also president of this company; and, again, Andrew W. and Richard B. Mellon were directors.[39]

Outwardly, these companies appeared to prosper during the next few years. Their development closely paralleled the growth of the Texas Company. The valuable Spindletop leases held by the Guffey company gave it an entry into the growing fuel oil market. It operated extensive field storage and collecting systems with a pipeline to terminal facilities at Port Arthur, where shipments were made to eastern markets via a company-owned fleet of ten tankers. As Spindletop production diminished, the Guffey company also expanded into the Sour Lake, Batson, and Saratoga fields. The Gulf Company's Port Arthur refinery, a much larger installation than the Texas Company's, manufactured marketable quantities of kerosene, as well as refined fuel oil and asphalt.[40]

Yet all was not well within the managerial structure of these companies. James M. Guffey, the restless old wildcatter who headed them, had interests elsewhere and never again visited Texas after the J. M. Guffey Petroleum Company was incorporated in May 1901.[41] Fortunately, the Mellons were no strangers to the oil business. They had operated a petroleum pipeline in the Pennsylvania field. Standard Oil had tried to prevent its construction but the Mellons persevered and, in the early 1890s, they sold the pipeline to Standard for a substantial profit. The Mellons dispatched to Texas a younger member of the family, William L., a nephew of Andrew and Richard. He returned with a report highly critical of Guffey's absentee management. To salvage their Texas investment, which by mid-1902 totaled over five million dollars, the Mellons forced Guffey to turn administrative control over to William L. Mellon, who was made executive vice-president of the Guffey-Gulf companies. He immediately employed a number of experienced oilmen, including several former employees of Standard Oil, to run

39. Craig Thompson, *Since Spindletop: A Human Story of Gulf's First Half-Century* (Pittsburgh: Gulf Oil Corporation, 1951), pp. 12, 14.

40. *Ibid.*, pp. 9–16, 17–22.

41. *Ibid.*, p. 13.

the companies; in time, young Mellon himself would develop into one of the country's finest oil executives. But at this point in his career, he was relatively inexperienced, especially with the Texas petroleum industry. The chaotic development of Spindletop and subsequent Gulf Coast fields left the young Easterner bewildered. "That Texas oil is such a headache," he is reported to have said repeatedly.[42]

Doubtless the Mellon family viewed the future of their Texas ventures with apprehension, since additional capital was required to save the initial investment. They even offered to sell the Texas properties to their old adversaries, the Standard Oil Company. Standard executive Henry H. Rogers quickly disabused the Mellons of this possibility. Recalling Standard's past problems with its marketing affiliate, Waters-Pierce, and Texas antitrust laws, Rogers claimed that John D. Rockefeller would "never put another dime in Texas."[43] Against this background of mounting problems, the Mellons eventually considered an obvious and logical possibility: why not join forces with an oil company also spawned by Spindletop, but one which could furnish proven leadership—Cullinan's Texas Company?

In early 1903, however, there appeared little chance of such a merger. The operation of adjoining oil properties at the Sour Lake field occasioned instances of bitter animosity between personnel of the Texas and Guffey companies. James R. Sharp, an officer of Producers Oil in charge of operating the Sour Lake lease, reported that Guffey Company employees frequently trespassed across a corner of the Texas Company property as they laid pipe to nearby storage tanks.[44] Repeated warnings brought no end to this practice, and Sharp concluded that the "Guffey Company were the meanest people on earth!"[45] The Texas Company subsequently employed armed guards to patrol its property.[46]

42. *Ibid.*, p. 22.
43. *Ibid.*, p. 18.
44. James R. Sharp (Sour Lake) to Cullinan (Beaumont), July 30, 1903, Cullinan Papers.
45. Sharp (Sour Lake) to Thomas A. Carlton (Pittsburgh), August 7, 1903, Cullinan Papers.
46. *Oil Investors' Journal*, October 1, 1903, p. 4.

Some months later, Cullinan personally played a key role in protection of the Texas Company's interests at Sour Lake. Word reached him that three of his employees had sold the Guffey Company engineering maps which indicated the proposed location of future drilling sites. Cullinan spent a week gathering conclusive evidence before confronting and discharging the guilty men.[47]

An earlier dispute concerned an oil contract between the two companies. Early production from the Guffey leases at Sour Lake was not impressive. The company needed oil and contracted with the Texas Company to purchase 7,000 barrels of crude oil daily over a period of three months. Later, the Guffey Company developed sizeable production at Sour Lake and, according to the Texas Company, it refused to accept and pay for the daily deliveries specified in the agreement. This was the last straw; Cullinan directed his lawyers to file suit for breach of contract. He wrote James M. Guffey that his company regretted taking this step but "our rights seem to be wholly ignored by your representatives at this end."[48]

Both Guffey and the Mellons now realized that the situation in Texas was getting out of hand. They proposed a meeting of Texas and Guffey company officials to discuss the disputed contract. Cullinan assented and it was agreed to begin the talks at Pittsburgh the first week of August 1903.[49] The Texas Company president, however, would be in Saratoga, New York, at that time—his first vacation since the organization of the company. He decided to send Walter B. Sharp, president of Producers Oil, the Texas Company's producing affiliate. It was an excellent choice: Sharp, a rangy, field-hardened six-footer with an incisive mind, met a congenial Andrew W. and William L. Mellon and soon surmised that the talks had implications beyond the mere settlement of an oil contract. If the Mellons' affability was bait for larger fish, Sharp still seized it to settle the immediate problem at hand. It was agreed to execute a new contract which would include delivery of an additional 300,-

47. Fred W. Freeman (Beaumont) to Arnold Schlaet (New York), July 28, 1904, Cullinan Papers.

48. Cullinan (Beaumont) to Guffey (Pittsburgh), July 1, 1903, Cullinan Papers.

49. Cullinan (Beaumont) to W. J. McKie (Corsicana), July 18, 1903, Cullinan Papers.

ooo barrels of crude oil; and the Mellons gave Sharp, on the spot, a check for $48,000 to square the Guffey Company's liability under the oil agreement. Sharp came back to Texas with a settlement which pleased Cullinan; he also relayed the news that the Mellons wanted to talk about the merger of the Guffey and Texas companies.[50]

Cullinan's first reaction to this news was one of curiosity and caution. He confided to his old Corsicana friend, W. J. McKie, that he had "no objection as to seeing what the Mellons had in mind," but that "after all these years of hard work we do not intend to give our properties away."[51] Although he did not mention his probable role in such a merged organization, McKie immediately grasped it by replying, "With JSC as manager of both concerns, all the oil in the field ought to be cornered!"[52]

Both Arnold Schlaet and John W. Gates agreed with Cullinan that there was no harm in talking with the Mellons, provided, of course, that any plan adopted would benefit the Texas Company's stockholders. Gates informed Cullinan of the problems he would probably encounter in talks with the Mellons. The financier pointed out that, since the assets of the Guffey-Gulf companies appeared larger than the Texas Company's, the Mellons would demand a proportionately larger share of the merged organization's stock. But Gates pointed out that the Texas Company could easily expand its physical facilities "out of earnings . . . and have as good a property as the Guffey Company and still continue to pay reasonable dividends." He further stated that "while I am not opposed to the consolidation . . . unless we could get a very much larger proportionate amount for our shares of stock than they are getting for theirs it would be a bad trade."[53]

Talks began in Pittsburgh on October 28, 1903. Cullinan and Schlaet represented the Texas Company; James M. Guffey, Andrew

50. Walter B. Sharp (Pittsburgh) to Cullinan (Saratoga, New York), August 14, 1903; Walter B. Sharp (Pittsburgh) to James R. Sharp (Sour Lake), August 14, 1903; Cullinan (Beaumont) to W. J. McKie (Corsicana), August 20, 1903, Cullinan Papers.

51. Cullinan (Beaumont) to McKie (Corsicana), August 28, 1903, Cullinan Papers.

52. McKie (Corsicana) to Cullinan (Beaumont), August 30, 1903, Cullinan Papers.

53. J. W. Gates (New York) to Cullinan (Pittsburgh), October 27, 1903, Cullinan Papers.

W. Mellon, and William L. Mellon represented the Guffey-Gulf companies. As Gates had prophesied, the Mellons wished to base stock ownership in a new organization on the proportionate share of assets contributed to the merger. Since the assets of the Texas Company were then about eight million dollars and the assets of the Guffey-Gulf companies about twelve million, the Pittsburgh bankers proposed that 40 percent of the stock in the new concern would be held by the Texas Company shareholders and 60 percent by the Mellon-Guffey interests. Cullinan countered that the assets of the Texas Company were undervalued and that the proportions of stock assigned each group should be at least equal. After two days of fruitless talks, Cullinan telegraphed Gates, "They are very anxious to trade but hold that our demands are unreasonable and think that if A. W. M. and yourself could meet some agreement might be reached."[54] If the Mellons had known of Gates's earlier advice to Cullinan, they could have spared themselves a short trip west. Nevertheless, it was arranged that Andrew and William Mellon, Calvin Guffey, nephew of James M. Guffey, and Cullinan would meet Gates the next week in Chicago.[55] But two more days of talk brought no agreement. Gates, of course, agreed with Cullinan and Schlaet that the Texas Company properties could not be merged on a 40-percent basis. It was decided to adjourn the discussion and give each side "a month or so to think over the other's proposition."[56]

Meanwhile, despite the efforts of both groups to keep the merger talks secret, news of the discussions leaked to the press. The Galveston *News* on October 27, 1903, stated that "it is rumored the J. M. Guffey Petroleum Company is soon to absorb the Texas Company."[57] John W. Gates sent Cullinan a clipping from the Chicago *Daily Trade Bulletin*, October 31, 1903, which announced that the merger was completed and that Gates as "the heaviest stock-

54. Cullinan (Pittsburgh) to Gates (New York), October 29, 1903 (telegram), Cullinan Papers.
55. Cullinan (Pittsburgh) to Gates (New York), October 30, 1903 (telegram), Cullinan Papers.
56. Cullinan (Beaumont) to W. J. McKie (Corsicana), November 9, 1903, Cullinan Papers.
57. Clipping in the Cullinan Papers.

holder in the Texas Company . . . received $25,000,000 for his interest." In an accompanying letter, Gates said, "I wish the part about the 25 million was true."[58]

There was, however, no rush to resume negotiations. Gates reported that Andrew Mellon, whom Gates happened to meet in a New York club, was unwilling to change his stand on the merger plans.[59] Gates saw no reason for concern. Time and the profitability of the Texas Company would bring the Mellons around. He later advised Cullinan:

It seems to me if we were to trade with the Guffey people now we would want a better basis than we had when you and they first opened negotiations last fall. Anyone who attempts to manage a property a thousand miles away from it always is at a disadvantage and I think the Mellons will eventually make up their minds that they are no exception.[60]

Cullinan agreed that discussion with the Mellons should be continued, for, as he told Gates, "I think that merging, if it can be done on safe and conservative lines, is undoubtedly the proper thing for both interests. . . . " But he pointed out that a major problem had arisen in the earlier talks. Cullinan felt the Mellons had failed to reveal sufficient detail concerning the value of their Texas properties. Further talks would be useless unless there was more information. "I can not see how we are to arrive at a comparative value of their property," he stated, "without going through their [the Mellons'] affairs from end to end."[61]

Cullinan reiterated this contention when the Mellons suggested the resumption of merger talks in the fall of 1904. This time, the Mellons agreed to an extensive audit of their Texas properties and a joint team of Texas and Guffey company accountants went

58. Gates (Chicago) to Cullinan (Beaumont), November 10 ,1903, Cullinan Papers.
59. Gates (New York) to Cullinan (Beaumont), January 6, 1904, Cullinan Papers.
60. Gates (New York) to Cullinan (Beaumont), February 17, 1904, Cullinan Papers.
61. Cullinan (Beaumont) to Gates (New York), May 12, 1904, Cullinan Papers. It is possible that the Mellons at this point did not know the actual value of the Guffey-Gulf properties. The J. M. Guffey Company was capitalized in 1901 at $18,000,000, which was "a bookkeeping valuation . . . based more on expectations than realistic appraisal. . . ." (Thompson, *Since Spindletop*, p. 12.) This source also notes (p. 95) that the value of the assets of these companies for the years 1902 through 1906 was "Not Available."

to work at Beaumont for this purpose. Later Arnold Schlaet represented the Texas Company as the final compilation of property schedules took place at Pittsburgh.[62] Ironically, when the assets of both companies, based on a fair-market value, had been totaled, it was found that the Mellons were correct in their previous appraisals. The assets of the Guffey-Gulf companies were $11,122,000; the assets of the Texas Company were $7,348,000—or 60.22 percent and 39.78 percent, respectively, of the combined total evaulation of $18,470,000.[63]

Arnold Schlaet immediately telegraphed the essence of the evaluation report to Texas and urged that merger talks be resumed.[64] Cullinan agreed and went to Pittsburgh for further discussions with the Mellons. After two weeks of negotiations, a preliminary plan of merger was adopted and signed on November 30, 1904, by Cullinan and William L. Mellon as president and vice-president of their respective companies. The plan proposed that a new company would be chartered to engage in the production, manufacture, and transportation of petroleum; and it would be formed from the properties held by the Guffey-Gulf and Texas companies. The concern was to be capitalized at $22,500,000; 60 percent of its stock was to be distributed to the Guffey-Gulf stockholders and 40 percent to the Texas Company stockholders. In addition, the Texas Company stockholders were to receive a liquidating dividend of $333,000 in cash. This amount was to be paid by the Mellons' Guffey-Gulf companies for the adjustment of current oil inventories and contracts. Within one year after incorporation, the sum of three million dollars was to be raised through sale of stock or issuance of bonds to furnish additional funds for investment in

62. Cullinan (Beaumont) to W. J. McKie (Corsicana), October 12, 1904, Cullinan Papers.

63. The components of the Guffey-Gulf evaluation of $11,122,000 were: pipelines and storage facilities, $2,354,000; refineries, $3,766,000; ships and railroad cars, $2,454,000; real estate, $1,047,000; and oil inventories, contracts, and properties, $1,501,000. The Texas Company's total assets of $7,348,000 were: pipelines and storage, $1,723,000; refineries, $1,018,000; ships and railroad cars, $235,000; real estate, $622,000; and oil inventories, contracts, and properties, $3,750,000. (Preliminary merger agreement, dated November 30, 1904, Texaco Archives, II, 175.)

64. Schlaet (Pittsburgh) to Cullinan (Beaumont), November 7, 1904 (telegram), Cullinan Papers.

the new enterprise. The terms of the merger were to be ratified by the stockholders of the companies involved, and the incorporation of the new company was to be in full compliance with Texas law.[65]

Although it was not included in the preliminary merger agreement, there was an informal meeting of minds at Pittsburgh concerning the management of the new company. Cullinan was to be its president, although a representative of the Mellons, probably William L. Mellon, would serve as one of the vice-presidents. It was also planned that Arnold Schlaet and Walter B. Sharp would eventually hold important posts in the new concern. Cullinan proposed to resign the presidency of the Texas Company in January 1905, and devote his time to supervising the merger details. Meanwhile, he suggested that Schlaet become president of the Texas Company "since the final planning of the merger is likely to stretch out for several months and Schlaet is fully qualified to run operations until then."[66]

The report of the Pittsburgh agreement won over John W. Gates, despite the financier's earlier reluctance to merge the Texas Company's properties for a mere 40 percent interest in the new firm. He again deferred to Cullinan and stated that he was "particularly pleased by your [Cullinan's] part in the new company . . . and it seems like a good trade."[67] Schlaet, who was present at the Pittsburgh discussions, kept the Laphams informed and won their approval of the preliminary agreement. Lewis H. Lapham later wrote Cullinan that the proposed merger was "an arrangement for the mutual advantage of all concerned."[68]

Cullinan came back from Pittsburgh satisfied that an agreement in the best interests of the Texas Company stockholders had been reached. He was also unquestionably flattered and pleased to be selected president of the new enterprise. Yet he was sobered by these new responsibilities. He wrote Schlaet that "I have prac-

65. Preliminary merger agreement, dated November 30, 1904, Texaco Archives, II, 175; copy also in Cullinan Papers.
66. Cullinan (Pittsburgh) to J. W. Gates (Chicago), October 15, 1904, Cullinan Papers.
67. Gates (Chicago) to Cullinan (Beaumont), November 10, 1904, Cullinan Papers.
68. Lapham (New York) to Cullinan (Beaumont), November 22, 1904, Cullinan Papers.

tically deserted my family for the past three years and devoted my entire time to business . . . and the new proposition will mean living in the same way for the next two or three years." For a moment there seemed a brief shadow of a doubt as he pondered the problem of managing an organization in which the Mellons would control a majority of the stock. "I fear, perhaps," he told Schlaet, "that you and I have possibly organized a mutual admiration society that may not appeal to them." Nevertheless, he felt certain that "the sound business practices we have exhibited in the operation of the Texas Company will make the consolidation successful."[69]

Meanwhile, the Texas Company's general attorney, James L. Autry, cautioned Cullinan that the new company's charter, which proposed to empower it with producing, transportation, and manufacturing functions, was incompatible with existing Texas law. He wrote:

It is a problem similar to that we faced upon incorporation of the Texas Company . . . for neither the Pipeline Act of 1899 nor the state's general incorporation laws permit a charter to be filed with more than one corporate purpose. Furthermore, we can expect state authorities, in view of the magnitude of the undertaking [merger of the Guffey-Texas companies], to examine the transaction very closely.[70]

This problem obviously could not be taken lightly and it apparently had been discussed earlier at the Pittsburgh meeting. Soon after the preliminary agreement had been reached, further discussions were held at Beaumont. Robert A. Greer and W. B. Markham, prominent local attorneys and directors of the Guffey-Gulf companies, represented the Mellons; Cullinan, Autry, and former Governor James S. Hogg represented the Texas Company.[71]

At these talks there was unanimity on a general course of action.

69. Cullinan (Beaumont) to Schlaet (New York), November 30, 1904, Cullinan Papers.

70. Autry (Beaumont) to Cullinan (Beaumont), November 21, 1904, Cullinan Papers.

71. Hogg was retained as an associate counsel to assist in the merger and was to be paid $5,000 for his services. (Cullinan [Beaumont] to Schlaet [New York], January 21, 1905 [telegram], Cullinan Papers.)

Since Texas law did not permit the incorporation of an integrated petroleum company, the law either had to be changed or a way found to make the proposed company a statutory exception. There was a difference of opinion, however, as to which of these stratagems to employ. Greer and Markham recommended that a bill be introduced in the current state legislature amending the Pipeline Act of 1899 to permit integrated operations. The Beaumont attorneys felt this to be the best plan because such legislation would not specifically allude to the forthcoming Guffey-Texas companies merger. Cullinan, on the other hand, suggested that the Texas legislature be requested to permit the incorporation of the integrated company by special enactment. The merger talks were no longer a secret and both Cullinan and Autry felt that if the whole matter was placed before the legislature in its true light it would bring approval for the incorporation of an integrated company. But after lengthy discussion, Cullinan and his legal advisors acceded to the wishes of the Mellon attorneys—particularly as Greer stated that he had already gained the support of Walter B. Myrick, Jefferson County (Beaumont) delegate to the state House of Representatives, who would introduce the amendment to the Pipeline Act of 1899.[72]

Representative Myrick thus introduced House Bill No. 324 in the Texas legislature on January 26, 1905. The bill provided that companies formed for the transportation and storing of petroleum would also have the right to prospect for, produce, and refine oil and gas, and, as under the Pipeline Act of 1899, such companies would retain the right of eminent domain. After its first reading, the bill was referred routinely to the House Committee on Private Corporations for hearings.[73]

The "Myrick Bill," however, was to meet determined opposition from the state's smaller petroleum producers. The pent-up bitterness and frustration of these operators, simmering the four years since Spindletop, now burst out in indignation against the bill.

72. James L. Autry (Beaumont) to Arnold Schlaet (New York), January 23, 1905, Cullinan Papers.

73. Texas, *House Journal, 29th Legislature, 1905* (Austin: State Printers, 1905), pp. 226–227, 254.

Moreover, at the same time the Myrick bill was being debated, the newspapers were filled with sensational disclosures concerning investigations of the Standard Oil Company by the Kansas attorney general. These investigations indicated that Standard Oil monopolized the Kansas petroleum industry through control of production and transportation facilities, and, in March 1905, state antitrust action was instituted.[74] In such an atmosphere, the spokesmen for the Texas operators rallied their forces to battle with cries that the Myrick bill would result in a "Texas petroleum trust" similar to that inflicted upon Kansas.[75]

To air their complaints further, these smaller Texas operators formed the Oil Producers' Association of Beaumont and sent a committee, headed by Beaumont attorney Robert A. John, to represent them at the Myrick bill hearings before the House Committee on Private Corporations. At these hearings, the producers' delegation told of alleged wrongs inflicted upon the small producers by the large transportation, storage, and refining companies that differed little from similar allegations made in the Pennsylvania fields a generation earlier. Conveniently forgetful of their own irresponsible contributions to the chaotic development of every Texas oil field since Corsicana, these producers told of forced sales of crude oil below market price to these larger companies, of excessive storage and pipeline charges inflicted upon them, and of the exorbitant profits subsequently earned by the larger companies at the expense of the smaller producer. In short, as the head of the producers' delegation, Robert A. John, pointed out, legislation such as the Myrick bill would compound these evils; what really was needed was legislation protecting the "helpless producer from pipeline and refining companies speculating in oil!"[76]

Meanwhile, Representative Myrick, claiming that the proposed legislation "was misunderstood . . . and had no concealed meaning," attempted to ameliorate the situation by offering an amendment to his original bill. The major premise of the original bill was re-

74. Hidy and Hidy, *Pioneering in Big Business*, pp. 671–676, 683.
75. Houston *Post*, February 4, 1905.
76. *Ibid.*, February 17, 1905.

tained—that is, that pipeline companies could engage in producing and manufacturing activities—but transportation and storage charges assessed by such companies against other producers would be subject to regulation by the state railway commission. This was an apparent concession to the smaller producers, for although the Pipeline Act of 1899 designated companies chartered under its provisions as "common carriers," it did not bestow regulatory powers upon the railroad commission. The act specified penalties if the pipeline companies discriminated against a producer in charges or service, but the complaint was heard and adjudged through usual court procedures. Myrick's amendment now would give the smaller producers access to the regulatory powers of the state railroad commission in the airing of their grievances.[77]

This apparent concession only encouraged the small operators to take a stronger stand. With the backing of the Oil Producers' Association, Representative Chester H. Bryan of Harris County introduced House Bill No. 484 on February 15, 1905. This bill not only excluded transportation companies from producing and manufacturing activities but made provision for pipeline and storage charges to be subject to statutory regulation by the state legislature. A companion bill was introduced into the Senate by D. E. Decker of the Twenty-Ninth, West Texas, district.[78]

At this point, Cullinan became quite concerned about this legislative maneuvering. He had agreed to give Robert A. Greer, the Mellons' Beaumont attorney, the responsibility of managing the passage of the Myrick bill. But already Cullinan was convinced that "Mr. Greer seems to have made a 'bust' on introducing his bill." He also confided to Arnold Schlaet:

> It is unfortunate that our original plans [i.e., asking the legislature to approve the Guffey-Texas Companies merger by special enactment] were not carried out, and the matter have been made public instead of being placed on the defensive, which is the present attitude.[79]

77. *Ibid.*, January 29, February 4, 1905.
78. Texas, *House Journal, 29th Legislature, 1905*, p. 412; Houston *Post*, February 16, 1905.
79. Cullinan (Beaumont) to Schlaet (New York), February 3, 1905, Cullinan Papers.

While Cullinan approved of the amended bill offered by Myrick, his prior consent had not been sought, and it was "out of line with what was agreed to as between the Mellon people and ourselves." Moreover, should the merger not go through, this tinkering with the Pipeline Act of 1899 stood to work for the future detriment of the Texas Company. That company was chartered under the act of 1899 but the Guffey Company, on the other hand, was chartered under Texas corporation law authorizing the formation of private petroleum transportation companies. Cullinan resolved that, win or lose, it was time to go to Austin to present matters in "an open, candid way" and protect both the interests of the Texas Company and "my own reputation."[80]

Both Cullinan and Autry made it a point to be present at the next public hearing on the Myrick bill conducted by the House Committee on Private Corporations. The session was held on the evening of February 16, 1905, in the ballroom of the Driskill Hotel in Austin. Cullinan soon had firsthand evidence that Robert Greer's handling of the Myrick bill was neither tactful nor likely to win over dissenters. Greer spoke in support of the bill and Robert A. John, head of the small producers' delegation, spoke in rebuttal. Hot words were exchanged and after the session adjourned, Greer and John came to blows in the hotel lobby. Before a "great crowd" which quickly gathered, Cullinan and Autry separated the combatants and prevented "a serious affray" from occurring. A newspaper justifiably commented that the Myrick bill was "about the warmest question which has come up this session of the legislature."[81]

The next public hearing on the Myrick bill was scheduled for February 24, 1905. Cullinan asked to testify and the House Committee quickly accepted his offer to appear. Committee members were reported to be very eager to ask him about the rumors concerning the merger of the Guffey-Texas companies and it was said that "another warm time was expected with Cullinan on the stand."[82]

80. *Ibid.*
81. Houston *Post,* February 17, 18, 1905.
82. *Ibid.,* February 21, 1905.

As he began his testimony before the committee, Cullinan read a prepared statement, which he had written, entitled "Facts Concerning the Oil Industry of Texas."[83] The document traced the major trends in the state's petroleum development from the Corsicana field to the present. It noted the annual production of the various Texas fields and the number of firms currently engaged in all phases of the industry. The oilman claimed that the Texas Company had always made uniform charges for services rendered to other producers and, further, that the fees charged for these services "in all the older fields of the north are 200 to 400 percent higher than what are charged in Texas." He pointed out that in 1904 the Texas Company made an average profit of 6.71 cents on each barrel of crude oil it sold. "While this is a business detail," he added, "it is given frankly for the purpose of refuting statements made before this committee that the pipeline companies received enormous profits as between the price at which they purchased and sold this commodity."

Cullinan also pointed out that the Texas petroleum industry, after four years of extensive development since the Spindletop discovery, now faced a severe crisis. It badly needed additional capital to compete with new petroleum-producing areas in Oklahoma, Kansas, and California. Moreover, the sizeable amount of capital required for industrial expansion meant that it would undoubtedly have to be supplied by non-Texas investors. (Here, he candidly cited his experiences with the organization of the Texas Company: "Probably one-third [of the capital] was owned by residents of this state; the balance represented eastern or northern capital.") This foreign capital could not be attracted into the state unless legislation permitted the formation of "compact organizations . . . concentrating the facilities, energy, and capital" to engage in all phases of the petroleum industry. These companies would bring stability to Texas oil development and insure that the landowner, producer, and manufacturer, alike, would receive a reasonable compensation for his product. The Texas Company president concluded his statement with a plea for the enactment of the "My-

83. This statement subsequently appeared in the *Oil Investors' Journal,* March 3, 1905, pp. 10–11. The following references are from this source.

rick Substitute Bill," which permitted integrated operations and yet gave the state railroad commission power to regulate pipeline and storage charges.

Representative Thomas B. Love of Dallas County, committee chairman, thanked Cullinan for his "open and masterful" statement but then asked about the Guffey-Texas companies' merger. Cullinan did not equivocate. He stated that "talks were under way which, if successful, will merge the two companies." The session was adjourned without further questions, but it was reported that Cullinan's "confession . . . made an adverse impression on the committee."[84] Indeed, the opponents of the Myrick bill were quick to use this news. Robert A. John, spokesman for the Oil Producers' Association, subsequently proclaimed that passage of the Myrick bill, in the light of the impending Guffey-Texas companies' merger, "would undoubtedly enable the great capitalists of the North to seize complete control of the Texas oil industry."[85]

Cullinan returned from Austin discouraged and convinced that the Myrick bill would not be enacted. To compound his dejection, it appeared that the Bryan bill might have a chance of passage. "We prefer to get no legislation," he wrote John W. Gates, "rather than to get this bill . . . which is so inimical to our interest."[86]

The Bryan bill had been introduced with the blessing of the small producers' delegation at Austin. It had two major features: it excluded pipeline companies from engaging in producing and manufacturing functions; and it gave the state legislature authority to set pipeline and storage charges by specific statutory enactment. Cullinan was, of course, disturbed about the first feature. The thinly concealed relationship between the Texas Company and producing affiliates such as the Producers Oil and Moonshine Oil companies might be subjected subsequently to official scrutiny.

The second feature especially incensed Cullinan. While he would allow the state railroad commission the authority to regulate pipeline rates, the thought of authorizing the state legislature to enact these rates appalled him. As many oilmen were to affirm after him,

84. Houston *Post*, February 28, 1905.
85. *Ibid.*, March 3, 1905.
86. Cullinan (Beaumont) to Gates (New York), March 8, 1905, Cullinan Papers.

Cullinan felt that the railroad commission, composed of a small, but select group of reasonable men, perhaps with experience in the oil industry, were better qualified to regulate petroleum matters than a large and varied legislative body molded by political pressures. "To permit regulation by the legislature," as he put it, "would allow the grafters a chance to get in their work."[87]

A vigorous counterattack against the Bryan bill was thus launched. Cullinan informed Arnold Schlaet that he had elicited the support of an old Corsicana friend, Representative Richard Mays, delegate from Navarro County, to oppose the bill. Furthermore, the influence of former Governor James S. Hogg, who had been retained as counsel to assist in the merger of the Guffey-Texas companies, would be helpful. Cullinan stated that "between Mays and Governor Hogg, I believe we are in a position to protect ourselves and kill this [the Bryan] bill. . . . "[88]

Whatever this team did in legislative infighting, it was successful. By March 1905, a newspaper reported that the supporters of the various pipeline bills had succeeded in confusing a situation which started as a "simple request to extend existing law." The same source predicted that with the legislature striving for an adjournment in early April, "there would be no pipeline bill passed this term for the House Committee [on Private Corporations] is hopelessly deadlocked as to which of the bills to endorse."[89] This was so. Unable to arrive at a decision, the committee on March 13, 1905, reported unfavorably on both the Myrick and Bryan bills. [90] Cullinan, with some relief, wrote Gates that there would be no pipeline legislation at that session. "I regret, of course, that the Myrick bill was not passed," he added, "but a delicate situation was mishandled from the beginning."[91]

While Cullinan was preoccupied with events in Austin, Arnold Schlaet periodically met with the Mellons in Pittsburgh to arrange

87. Cullinan (Beaumont) to Arnold Schlaet (New York), February 3, 1905, Cullinan Papers.

88. *Ibid.*

89. Houston *Post,* March 8, 1905.

90. Texas, *House Journal, 29th Legislature, 1905,* pp. 735, 755.

91. Cullinan (Beaumont) to John W. Gates (New York), March 18, 1905, Cullinan Papers.

further details of the merger. A disagreement arose over the payment of the expenses of the consolidation, which it was estimated
would be $75,000. Schlaet claimed that the Mellons initially agreed
to bear this entire expense. Later, they changed their minds and demanded that the expense be shared by the two groups, as the stock
in the new enterprise would be shared, on a "60–40" basis. Schlaet
was particularly irritated by the reason the Mellons gave for their
change of mind. They told Schlaet that the other stockholders of
the Guffey Company objected to that company's assumption of all
the merger expenses. Recalling that the Mellons were the majority
stockholders in that company, Schlaet felt that such reasoning "was
trivial," and as far as he was concerned, "our deal with the Mellons
is . . . off."[92]

Early in February 1905, Schlaet was advised to "hold the matter
of the [merger] fees in abeyance until matters in Austin are concluded."[93] But now the refusal of the state legislature to permit
integrated operations by the passage of the Myrick bill was obviously the deathblow to the merger plans. Cullinan informed Schlaet
of this in a letter touched with both disappointment and relief:
"Taken all in all the more I see of the way things were handled . . .
the more fortunate I feel we are in not having made a deal on the
lines proposed and merged with a family whose ideas were so widely
at variance to do things as ours and theirs."[94]

And to John W. Gates Cullinan confided that his "only regret
was that the merger talks took months of time that should have
been given to other matters."[95] Gates, sensing that Cullinan's lost
opportunity to lead the merged Guffey-Texas companies was still
a keen personal disappointment, resisted an "I-told-you-so" attitude. He replied that "I assure you that your devotion to the
[Texas] company has been appreciated and I look forward to con-
tinuing that relationship."[96]

92. Schlaet (New York) to Cullinan (Beaumont), January 27, 1905, Cullinan Papers.
It was noted that copies of this letter were sent to John W. Gates and Lewis H.
Lapham.
93. Cullinan (Beaumont) to Schlaet (New York), February 3, 1905, Cullinan Papers.
94. Cullinan (Beaumont) to Schlaet (New York), March 3, 1905, Cullinan Papers.
95. Cullinan (Beaumont) to Gates (New York), March 18, 1905, Cullinan Papers.
96. Gates (Chicago) to Cullinan (Beaumont), March 25, 1905, Cullinan Papers.

Meanwhile, the development of other Gulf Coast oil fields gave Cullinan little time to ponder the failure of the Guffey-Texas companies' merger. The Humble field, twenty miles north of Houston, reached its peak in 1905, with the production of 15,594,310 barrels of crude oil. To acquire this flush production, which dropped Gulf Coast crude oil prices to an average of twenty-four cents a barrel for that year, the Texas Company quickly constructed extensive storage facilities and built a fifty-mile extension of its Sour Lake-Port Arthur pipeline to the new field.[97]

To finance these improvements, the company raised its capitalization from the original $3,000,000 to $6,000,000 through the issuance of 30,000 new shares of stock. The company stockholders purchased 19,800 shares of this issue and the remaining 10,200 shares were quickly sold through underwriting agreements with New York and Chicago brokerage houses. Sale of the entire issue had been completed by April 30, 1906. There were now 237 Texas Company stockholders.[98]

But the rapid development and loss of flush production of the Humble field signaled a general decline of the entire Gulf Coast petroleum industry. In 1906, production at Humble slumped to 3,571,445 barrels and, with older fields such as Spindletop, Sour Lake, Batson, and Saratoga having passed their flush phases, total Gulf Coast production dropped to about 20,500,000 barrels after a high of 36,500,000 barrels the preceding year. The decline was even greater in the succeeding years: 16,400,000 barrels in 1907; 15,800,000 barrels in 1908; and 10,900,000 barrels in 1909. Meanwhile, new discoveries in the mid-continent area, particularly in the Indian Territory (Oklahoma) and in California, easily surpassed the Gulf Coast production. The mid-continent field's crude production was only 12,500,000 barrels in 1905, but it increased to 22,800,000 barrels in 1906, 46,800,000 barrels in 1907, 48,800,000 barrels in 1908, and 50,800,000 barrels in 1909. California production showed steady annual increases from 33,400,000 barrels in

97. Warner, *Texas Oil and Gas*, p. 372; Williamson, *et al., The American Petroleum Industry, 1899–1959*, p. 39; Burt E. Hull Memo, dated November 10, 1951, Texaco Archives, I, 43.
98. Texas Company Folder, Autry Papers.

1905 to 55,500,000 barrels in 1909. As a result of the increased productivity of these new areas and declining output in the Gulf Coast fields, the latter, which had produced 21.52 percent of the country's crude oil in 1904, produced only 5.94 percent in 1909.[99]

Cullinan was fully aware of what this decrease in Texas petroleum production meant to the future of his company. Consequently, he welcomed the news in early 1905 of the prolific Glenn Pool field discovery, near Tulsa in the Indian Territory, almost 550 miles miles north of Beaumont. He was particularly interested to learn that the Glenn Pool oil was of a very high grade. It rated from 32 to 38 degrees on the Baume specific gravity scale, with a paraffin base, and was far more adaptable to refining than the Gulf Coast crude oil. Furthermore, Glenn Pool oil was relatively cheap. Rapid exploration had brought the usual glut of production, and local producers, particularly handicapped by inadequate railroad transportation facilities, were forced to sell this high-grade oil at prices as low as thirty cents a barrel.[100]

Cullinan's old friend and legal adviser, W. J. McKie of Corsicana, represented Texas oilmen who had acquired oil properties in the Glenn Pool field. McKie tried to interest the Texas Company in these properties; he wrote Cullinan that "the Indian Territory development will offer your company great opportunities."[101] Cullinan agreed, but, at that time, the merger with the Guffey Company was a possibility and the Humble field was in its flush phase. "This has made," he stated, "the Eastern people in our company timid on taking up anything new."[102]

That attitude changed after the merger plans were dropped and Texas production steadily decreased in the wake of the rapid exploitation of the Humble field. Cullinan wrote John W. Gates and Arnold Schlaet for approval of plans to construct a pipeline connecting the refinery at Port Arthur with the Glenn Pool area. He insisted that "we must not remain aloof from the Indian Terri-

99. Warner, *Texas Oil and Gas,* p. 372; Williamson, *et al., The American Petroleum Industry, 1899–1959,* pp. 16–17.

100. Rister, *Oil! Titan of the Southwest,* p. 91.

101. McKie (Corsicana) to Cullinan (Beaumont), January 2, 1905, Cullinan Papers.

102. Cullinan (Beaumont) to McKie (Corsicana), January 18, 1905, Cullinan Papers.

tory and Oklahoma activity . . . and should begin construction [of the pipeline] as soon as possible."[103]

Furthermore, Cullinan knew that the Guffey Company was planning a similar pipeline. In fact, during the merger talks, the advisability of such construction had been discussed and, had the consolidation been completed, it would have been the first major project of the new company.[104] After the merger plans were dropped, however, the Mellons decided to reorganize their Texas properties. In January 1907, the J. M. Guffey Petroleum Company was dissolved and a new producing company, the Gulf Oil Corporation, was formed. The Mellons overrode the remaining interest of the old wildcatter, James M. Guffey—he later complained that "they throwed me out"[105]—and solidified their control of both the new company and its manufacturing subsidiary, the Gulf Refining Company. Andrew W. Mellon was elected president of both companies and another subsidiary was formed, the Gulf Pipe Line Company, to construct the pipeline from Port Arthur to the Glenn Pool field.[106] Plainly, the Mellons were in the southwestern oil industry to stay. Their decision to build an Oklahoma pipeline now put further pressure on the Texas Company to construct a similar project.

On October 17, 1906, the directors of the Texas Company held a special meeting at Beaumont and approved Cullinan's plans to construct an Oklahoma pipeline. To pay for this project, the authorized capital of the Texas Company was increased from $6,000,-000 to $12,000,000 and an additional 60,000 shares of $100-par stock were issued. By June 1907, 20,000 shares of this stock had been sold; 50,000 shares were sold by June 1908. The remaining

103. Cullinan (Beaumont) to Gates (New York) and Schlaet (New York), August 6, 1906, Cullinan Papers.

104. Cullinan (Beaumont) to Gates (Chicago), January 2, 1905, Cullinan Papers.

105. Thompson, *Since Spindletop*, p.24. Guffey subsequently instituted suit against Andrew W. Mellon for $142,000 which he claimed was due him as reimbursement for expenditures he had made while president of the J. M. Guffey Petroleum Company. The case dragged through the courts for many years and was finally decided adversely for Guffey in 1926. He received nothing and, a few years later, the impoverished "old man died in a house which he fondly believed to be his own." (O'Connor, *Mellon's Millions*, p. 105.)

106. Thompson, *Since Spindletop*, pp. 23–26.

10,000 shares of the issue were paid to the company stockholders in 1909 as a stock dividend. Since 30,000 of these shares were offered for sale through brokerages, the number of Texas Company stockholders rose to 587 by 1909.[107]

With adequate financing assured, construction of the Oklahoma pipeline began in February 1907, and the entire eight-inch main trunk line system was completed by January 1908. The first Glenn Pool crude, flowing 560 miles, was delivered to the company's Port Arthur refinery in early February 1908. The system cost $5,800,000 and could deliver 17,000 barrels of crude oil per day.[108]

The route selected for the system indicated that company officials had planned with future marketing needs in mind. It started at the Glenn Pool area, near Tulsa, ran south into Texas through the small cities of Sherman and McKinney, and curved to the southwest through the major metropolitan area of Fort Worth-Dallas. The line then continued to the southeast through Corsicana and made connections with the existing company system at Humble. A twenty-mile extension to the south was soon added, giving the system a further access to tidewater via the Houston Ship Channel.[109]

With the construction of the Oklahoma pipeline, the Texas Company entered a new phase of product diversification and market expansion. Even before the pipeline was completed to Port Arthur, a refinery was constructed in the fall of 1906 at West Dallas to process Glenn Pool crude oil. At first that plant concentrated on the refining of fuel oil, but as the manufacturing adaptability of Oklahoma crude became increasingly evident, both the West Dallas and Port Arthur refineries met the demands of a growing market with a large number of diverse petroleum products. In 1907, at the Dallas State Fair, the company exhibited samples of forty man-

107. Texas Company Folder, Autry Papers.

108. *Oil Investors' Journal,* February 19, 1908, p. 10.

109. *Ibid.* The construction of the Gulf Company's Oklahoma pipeline also began in February 1907. Since a more direct north-south route was used between Port Arthur and Tulsa, it cost $4,800,000 and was 454 miles in length. It was operative by October 1907. The Gulf line, however, did not pass through a number of urban areas, as did the Texas Company system. (*Oil Investors' Journal,* February 19, 1908, p. 10.)

ufactured products. The exhibit included illuminating oils, household naphthas, "high-test gasoline suitable for all gas machines, soldering machines, racing automobiles and boats," lubricating oils, cylinder stocks, "special lubricants" such as automobile, harvester, threadcutting, and saw oils, cattle-dipping oil, road oil, and asphalt paint and roofing materials.[110]

While the use of Oklahoma crude and the construction of the West Dallas refinery enlarged the company's fuel oil market to include urban areas of central and northern Texas, this growing list of manufactured products necessitated additional distribution stations. This was particularly the case as the company, in the years 1907 to 1911, aggressively pushed nationwide retail sales of "Familylite," a kerosene-base illuminating oil. An officer of the company later explained that the Texas Company soon won a substantial share of the illuminating oil market by selling its high-quality products at competitive prices, by instructing distribution-station managers to resupply local retailers only upon demand, thereby relieving the retailer of storage and inventory burdens, and by offering its products as an alternative to those of the Standard Oil "Trust," particularly in the Southwest, where state disclosures of Waters-Pierce marketing tactics resulted in public revulsion against that company's products. Once the illuminating oil market was established, acceptance of other Texas Company products followed.[111] By the end of 1910, the company operated 11 large storage terminals and 229 product-distributing stations, including 136 in Texas, 12 in Mississippi, 13 in Louisiana, 12 in Oklahoma, 21 in Missouri, 8 in New Mexico, and 27 scattered throughout the eastern states of New York, Pennsylvania, Massachusetts, and New Jersey. The trade name *Texaco* was copyrighted in 1907, and, with the familiar green *T* superimposed upon a red star, became the hallmark of the company's products. The company was well aware of the value of advertising in a highly competitive market. In the

110. *Oil Investors' Journal*, November 5, 1907, pp. 20–21. Although the Texas Company began manufacturing gasoline for automobile fuel at the West Dallas refinery in 1906, gasoline would not surpass illuminating oil as the company's major refined product until 1913. (Refinery Records, Statistical Folder, Texaco Archives.)

111. Undated memorandum by Louis W. Kemp, Texas Company Operations Folder, Kemp Papers.

first half of 1910, it spent the then substantial sum of $16,300 in advertising its "Familylite" brand of illuminating oil alone.[112]

In this expansion of the company's marketing system, Arnold Schlaet again played an important role. His long association with the Lapham's United States Leather Company had given him wide experience in product distribution and sales. He unquestionably had more knowledge of these problems than did Cullinan, who had little direct experience with large-scale marketing operations. As a result, it was Schlaet who often originated policies concerning the location of bulk stations, the selection of local distributors, pricing, and product diversification. Cullinan realized the value of Schlaet's experience and usually quickly approved his suggestions. Cullinan, proud of his company's growth and expanding sales market, confided to W. J. McKie that "much credit for all this goes to Arnold Schlaet in New York."[113]

During this period, however, Cullinan did not relinquish his right of ultimate authority or slacken in his vigorous supervision of company affairs. The construction of the Oklahoma pipeline was, of course, a project to excite any oilman. Characteristically, Cullinan was on the scene. Both he and Autry took temporary offices in Dallas in the fall of 1907, to be closer to the nearby pipeline and refinery construction.[114] Cullinan also insisted that the main office of the Texas Company be moved from Beaumont to Houston in 1908. The movement of the Gulf Coast petroleum development in a westwardly direction away from Beaumont and Houston's access to tidewater as a terminal within the company's pipeline system made such a move practical. Cullinan, as early as 1905, prophesied "that the time will come—perhaps at no distant day—when we will want our General Office in Houston instead of Beaumont, as . . . Houston seems to me to be the coming center of the oil business for the Southwest."[115]

As the Texas Company entered its tenth year in 1911, Cullinan

112. *Ibid.;* Burt E. Hull Memo, dated November 10, 1951, Texaco Archives, I, 45.
113. Cullinan (Houston) to McKie (Corsicana), July 18, 1909, Cullinan Papers.
114. Burt E. Hull Memo, dated November 10, 1951, Texaco Archives, I, 45.
115. Cullinan (Beaumont) to Arnold Schlaet (New York), February 3, 1905, Cullinan Papers.

could well afford a feeling of pride and satisfaction in both his own and his company's accomplishments. From its beginning as a small crude oil transportation and purchasing concern, at first serving only a limited fuel oil market, the Texas Company now had a strong start toward a national distribution and marketing system for its diversified petroleum products. Founded with capital and assets of less than $2,000,000, the Texas Company had increased its capitalization to $27,000,000 and its assets to over $32,000,000.[116] It operated more than a thousand miles of trunk pipeline and field gathering systems and had constructed storage facilities for more than twenty million barrels of oil. The company owned three refineries: its first, at Port Arthur, constructed in 1903; the West Dallas refinery, built in 1907; and an asphalt processing plant at Port Neches, Texas, three miles north of Port Arthur, purchased in 1908. Terminal facilities were maintained at twelve Gulf and East Coast ports; European marketing and distribution were in progress through a storage plant at Antwerp, Belgium. The company owned over 1,000 tank cars, 5 locomotives, 6 ocean-going tankers, and more than 20 barges and lighters.[117]

In addition to this impressive physical plant, Cullinan had assembled a staff of trained and experienced personnel to furnish continuity of expert management for the ensuing years.

But not all of Cullinan's problems were successfully resolved during this period. Long and arduous work to merge the Texas Company with the Mellons' Guffey and Gulf companies floundered in an abortive attempt to change state laws to permit integrated petroleum operations. Nevertheless, under a firm hand, the company expanded and prospered. In 1911, that very success posed a major problem that would have to be faced shortly: had the demands of the complex operations and structure of the Texas Company now outmoded the domination of Cullinan's individual leadership?

116. On June 30, 1910, capitalization was raised from $18,000,000 to $27,000,000 by the payment of a 50 percent stock dividend. (Texas Company Folder, Autry Papers.)

117. Inventory of Texas Company Assets, dated June 30, 1910, Texas Company Folder, Autry Papers.

8. Controversy, Resignation, and New Ventures, 1911-1937

AFTER NINE YEARS of steady development and growth, the Texas Company suffered economic reverses in 1911 and 1912. The company's net income, which had been an all-time high of $6,424,189 for the year ending June 30, 1910, plunged to $2,574,361 for the fiscal year 1911 and to $2,201,069 for 1912. The company did not reduce dividends during these years and the decline in earnings was temporary, for net income increased to a new high of $6,617,729 in 1913.[1] But this setback naturally subjected the company's management to examination and reappraisal.

Cullinan, however, was not unduly disturbed by these reversals. The poor showings in 1911 and 1912 were disappointing, but heavy expenditures in those years were necessary to acquire new properties in the booming Electra field of north Texas and the Caddo field of northwestern Louisiana.[2] These holdings, Cullinan predicted,

1. Statistical Folder, Texaco Archives.
2. The Texas Company's producing affiliate, Producers Oil Company, drilled the

would become "very valuable . . . and worth several millions more than are represented in our accounts."[3] It was not a time for pessimism: it was just the opposite, for opportunity abounded. Cullinan had written Arnold Schlaet earlier that for less than $250,000 the company could acquire thousands of acres of abandoned leases in the Humble and Batson fields and by deeper drilling and reworking make these properties again productive. He urged Schlaet to raise eastern capital, resorting to short-term borrowing, if necessary, to finance this project.[4]

Cullinan's plans for continued expansion made little impression on Schlaet, now turned more conservative by the poor showing of the Texas Company and by the recent trend of developments within the country's petroleum industry. The usual cycle of exploitation in the new north Texas fields and the general uncertainty following the court-ordered dissolution of the Standard Oil Company[5] had made the eastern business community confused and cautious. Schlaet felt that for the present it was time to go slow, to "fit expenditures to income . . . and abstain from having a look at everything that comes along."[6] When Cullinan remonstrated, Schlaet peevishly snapped that unless this was done "it would be up to the South to raise any further funds."[7]

While policy differences were thus intensified by the Texas Com-

---

Electra field discovery well in 1911. This field soon reached its flush phase with 4,227,104 barrels of crude oil produced in 1912 and 8,027,154 barrels produced in 1913. (Warner, *Texas Oil and Gas*, pp. 229, 394.) The Texas Company completed a 250-mile trunk pipeline from Port Arthur to the Caddo, Louisiana, field in late 1911. (Burt E. Hull Memo, dated November 10, 1951, Texaco Archives, I, 53.)

3. Cullinan (Houston) to Arnold Schlaet (New York), March 11, 1913, Texas Company Folder, Autry Papers.

4. Cullinan (Houston) to Schlaet (New York), July 7, 1912, Texas Company Folder, Autry Papers.

5. The decision finding the Standard Oil Company (New Jersey) guilty of violation of the Sherman Antitrust Act was announced by the United States Supreme Court on May 15, 1911. The company was given six months to divest itself of its interest in thirty-seven affiliated companies. (Hidy and Hidy, *Pioneering in Big Business*, pp. 709–710.)

6. Arnold Schlaet (New York) to John J. Mitchell (Chicago), May 5, 1913, Texas Company Folder, Autry Papers.

7. Schlaet (New York) to Cullinan (Houston), July 22, 1913, Texas Company Folder, Autry Papers.

pany's problems in 1911 and 1912, Schlaet had shown previously
a growing resentment of Cullinan's managerial practices. In 1909,
with the company outwardly thriving after the completion of the
Oklahoma pipeline and with an expanding national market for its
products, Schlaet still felt all was not well and bitterly complained
about the situation to his old benefactor, Lewis H. Lapham.
Schlaet's letter to Lapham merits examination in full because of
its candid style and colorful description of the problems faced by
an expanding company, initially dominated by the centralized
authority of individual management but now undergoing struc-
tural and geographical diversification:

> I have encountered some very unpleasant experiences with JSC chief-
> ly because he thinks he knows everything and must butt into everything.
> The business has grown so big that to attempt to inform anyone of all
> points of contact is useless, and he is so persuaded of the importance of
> the Southern organization (which I do not deny) that he looks upon us
> here in New York as the tail of the dog, and a very small tail at that. He
> also has a decided opinion as to the merits or otherwise of certain of
> our northern employees, in which my judgment and his differ radically,
> and this is a matter in which he will have to give way, as I propose to
> pick my own lieutenants and expect him to accept them gracefully no
> matter what his own opinion of them may be, the same as I do with
> many of the appointments that he makes in the South and of which I
> do not approve.
> So far as JSC is concerned, I think he is working too hard, though
> physically he does not show it. He assumes in his correspondence of
> late frequently a very dictatorial tone which is disagreeable. I might not
> mind it so much if it did not indicate that it might sooner or later make
> for an impossible working together along present lines.[8]

Whether Cullinan knew of Schlaet's animosity at that time is
unknown. But when both Lewis Lapham and John W. Gates approached him before the annual Texas Company stockholders
meeting in the fall of 1909, he assented to their proposal to change
the company bylaws to permit the naming of an Executive Committee composed of four directors with the company president

---

8. Schlaet (New York) to Lewis H. Lapham (Rome, Italy), March 24, 1909, Texaco
Archives, XLIV, 209.

serving as chairman. He confided to Judge McKie that he "heartily approved of the idea . . . as it will undoubtedly lift some of the burden I have carried alone these past years."[9] Accordingly, company stockholders and directors approved the change in bylaws in November 1909, and, in addition to Cullinan, elected Gates, his close associate, Chicago banker-broker John F. Harris, Lewis Lapham, and Arnold Schlaet to the Executive Committee. At the same meeting, the number of company directors was increased to thirteen. In addition to Cullinan, Gates, Schlaet, Harris, and Lapham, the other eight directors named were employee-officers of the Texas Company: R. E. Brooks, James L. Autry, Elgood C. Lufkin, Thomas J. Donoghue, Martin Moran, Ralph C. Holmes, Clarence P. Dodge, and George L. Noble.[10]

Yet the formation of the Executive Committee did not bring peace and harmony to the company; instead, it brought further managerial discord. Cullinan now felt a sense of betrayal. In the ensuing deliberations of the Executive Committee, Gates usually sided with Cullinan, but Gates's close associate, John F. Harris, usually voted with Schlaet and Lapham. Within a year, Cullinan had threatened to resign from the company if Harris was not removed from the committee.[11] Gates, however, mollified Cullinan and a compromise was arranged. Harris remained but the committee was enlarged by the addition of two other directors. The directors selected were Elgood C. Lufkin, a company vice-president then assisting Arnold Schlaet in New York, and John J. Mitchell, president of the Illinois Trust and Savings Bank of Chicago, also a Gates associate, who replaced director Martin Moran.[12]

Still, Cullinan was not pleased with the committee. The frequent meetings, usually held in New York, were time-consuming and

9. Cullinan (Houston) to William J. McKie (Corsicana), September 30, 1909, Cullinan Papers.

10. Excerpts, Board of Directors' Meeting, November 16, 1909, Texaco Archives, XLIII, 173.

11. Cullinan (Houston) to John W. Gates (New York), October 24, 1910, Texaco Archives, XLIV, 210.

12. Gates (New York) to Cullinan (Houston), November 1, 1910, Texaco Archives, XLIV, 207; Excerpts, Board of Directors' Meeting, November 22, 1910, Texaco Archives, XLIV, 211.

useless since the other members of the executive committee assumed none of the responsibility for putting their ideas into effect.[13]
Moreover, Arnold Schlaet's conservatism, which stemmed from
the company's economic reversals of 1911 and 1912, infected the
other committee members. Cullinan bitterly complained "that for
the past year or more practically everything of importance that we
have put up to New York or recommended from this end has
either been objected to or turned down. . . . "[14]

As Cullinan struggled to maintain managerial control of the
Texas Company, he had to do battle without the support and advice
of John W. Gates. Gates died in Paris, France, on August 9, 1911,
at the age of fifty-six.[15] Cullinan immediately proposed that Gates's
son and sole heir, Charles G. Gates, take his father's place on the
Board of Directors and the Executive Committee.[16] This was done,
but Charles apparently had little inclination for the complexities
of corporate operation. At first, he attended a few meetings, but
other activities, particularly travel, held a greater attraction, and
he soon lost interest in the problems of the Texas Company.[17] He
was not long to survive his father. Charles G. Gates, only thirty-two,
died suddenly in New York City of a stroke on October 28, 1913.[18]

Another problem Cullinan faced as the showdown for managerial control approached was the overwhelming preponderance of
the Texas Company's stock now held by eastern investors. In 1902,
when the company was formed, one-third of the initial capital was

13. Cullinan (Houston) to Lewis H. Lapham (New York), October 20, 1911, Texaco
Archives, XLIV, 211.

14. Cullinan (Houston) to Elgood C. Lufkin (New York), July 17, 1912, Texas Company Folder, Autry Papers.

15. Wendt and Kogan, *John W. Gates*, p. 349.

16. Cullinan (Houston) to Lewis H. Lapham (New York), October 20, 1911, Texaco
Archives, XLIV, 211.

17. Even after Charles's interest in Texas Company affairs waned, Cullinan continued to correspond with him and frequently offered fatherly advice and comments
on general economic conditions. In one letter, written while Cullinan was attempting
to persuade eastern members of the Executive Committee to invest more capital in
Texas petroleum operations, he still advised Charles Gates that, since national
economic conditions were uncertain, it would be best "to play close to the shore for
some little time." Cullinan (Houston) to Charles G. Gates (Colorado Springs, Colorado), June 21, 1912, Texas Company Folder, Autry Papers.

18. Wendt and Kogan, *John W. Gates*, p. 354.

contributed by Texas investors. But beginning with the Sour Lake tract purchase in 1903, additional capital was supplied almost exclusively by the company's eastern stockholders, particularly by the Laphams and John W. Gates. As a result, the interest held by Texas stockholders steadily shrank to a smaller and lesser portion of the company's total capitalization.

In 1910, this domination by eastern capital was further increased as the Texas Company's stock was granted a listing on the New York Stock Exchange. At that time, the company's authorized capital was increased from $18,000,000 to $36,000,000 and an additional 180,000 shares of $100-par stock were to be offered for public sale.[19] By November 1913, the company had issued and sold 120,000 of these new shares, making a combined total of 300,000 shares outstanding and held by 1,637 stockholders. The first investors in the company still had large holdings: Mrs. John W. Gates, executrix of her husband's estate, held 25,034 shares, and 17,164 shares were still registered to Charles G. Gates, who had died just a month before; various members of the Lapham family held 31,253 shares, in addition to 2,015 shares registered to Arnold Schlaet; and Cullinan owned 7,134 shares. However, in the three years that the company's stock had been listed on the New York Exchange and thus subjected to public trading, new investors had acquired substantial holdings. Company director John F. Harris held 5,200 shares and an additional 8,472 shares were registered to his brokerage house, Harris, Winthrop and Company. James N. Hill, son of the railroad magnate, owned 6,288 shares. Eleven New York investment banks and brokerages, including J. S. Bache and Company, August Belmont and Company, Shearson, Hammill and Company, and Raymond, Pynchon and Company, held a combined total of 51,820 shares. After examination of the stockholders' list as of November 15, 1913, James L. Autry concluded that 224,405 shares of the Texas Company's stock were owned by eastern investors, while only 43,887 shares were held by Texas investors.[20]

19. Application of the Texas Company to the Committee on Stock List, New York Exchange, September 13, 1910, Cullinan Papers.

20. List of Texas Company Stockholders, November 15, 1913, Texas Company Folder, Autry Papers.

Meanwhile, disagreement among the members of the Executive Committee had increased and Cullinan resolved to make a stand at the next stockholder's meeting, to be held in Houston on November 24, 1913. Two weeks before the meeting, Cullinan, James L. Autry, and Will C. Hogg[21] solicited the proxy support of the Texas Company stockholders. An accompanying statement pledged that should a majority of the stockholders designate these three as their agents, the re-election of the company's Executive Committee would be indefinitely postponed.[22]

Three days later, the company's eastern investors struck back by sending stockholders a second proxy request. The request was signed by John J. Mitchell, who was noted as the president of the Illinois Trust and Savings Bank; Alonzo B. Hepburn, chairman of New York's Chase National Bank, who was elected a Texas Company director in 1911; James N. Hill, son of railroad magnate James J. Hill, and a director of the Northern Pacific Railway, who was proposed by the eastern group to fill the director's vacancy left by the recent death of Charles G. Gates; and company directors Arnold Schlaet and Lewis H. Lapham. The proxy request stated that this group was not in accord with Cullinan's plan concerning the postponement of the re-election of the Executive Committee. If this group prevailed, it promised to vote substantially for the same directors, officers, and executive committee.[23]

With the issue joined and the proxy battle under way, Cullinan vigorously entered the fray against heavy odds. In a further statement addressed to the stockholders, he made a strong appeal to sectional sentiment. He briefly reviewed the history of the company

21. Will C. Hogg, eldest son of former Governor James S. Hogg, who died in March 1906, was elected a director on January 6, 1913. Cullinan insisted that this would give "the old Texas crowd some recognition in company affairs." (Cullinan [Houston] to Lewis H. Lapham [New York], December 10, 1912, Cullinan Papers.) However, stock records of November 15, 1913, show only 245 shares registered to the Hogg family: 35 shares to Will and 210 to his sister, Ima. (List of Texas Company Stockholders, November 15, 1913, Texas Company Folder, Autry Papers.)

22. Proxy statement of J. S. Cullinan, James L. Autry, and William C. Hogg, dated November 10, 1913, Texaco Archives, XLVIII, 10.

23. Proxy Statement of John J. Mitchell, Alonzo B. Hepburn, James N. Hill, Arnold Schlaet, and Lewis H. Lapham, dated November 13, 1913, Texaco Archives, XLVIII, 14.

and asserted that "its original management, its corporate attitude and activities were branded with the name *Texas* and Texas ideals" and that its "headquarters and governing authorities should be kept and maintained in Texas." Thus, the Executive Committee, "of whom six out of seven live outside of the State," had undertaken to "usurp the legitimate functions of the Board (of Directors) . . . and company officers."[24]

Sensing the tide running against him, Cullinan made a direct appeal to Mrs. John W. Gates for her proxy support. He hoped that she would not "cast her influence on the side of the bankers and others having a small interest in the company relative to their whole means" against those "who have invested the bulk of their means . . . and for years given the best that there is in them to the company's upbuilding and development." It was a situation similar to those "your husband had to meet and combat during his entire business career."[25] But Cullinan received no acknowledgment from Mrs. Gates, and he glumly confided to Autry that "she probably will not help us."[26]

As the stockholders' meeting approached, the proxy contest was further complicated by rumors and innuendo sweeping New York and Houston financial circles. On a trip to New York earlier in the year, Cullinan called upon his old Corsicana benefactor, Henry C. Folger, Jr., at the Standard Oil Company offices located at the well-known address, 26 Broadway. Although the call was short and of a social nature, Cullinan was apparently seen leaving by a member of the opposing proxy committee.[27] It was now rumored that Cullinan was making a deal with the Standard Oil people. He supposedly

24. Statement to the Stockholders of the Texas Company by J. S. Cullinan, president, November 17, 1913, Texaco Archives, XLVIII, 15–16.

25. Cullinan (Houston) to Mrs. John W. Gates (New York), November 17, 1913 (telegram), Cullinan Papers. In his approach to Mrs. Gates, Cullinan might have done well to wave the Lone Star flag as enthusiastically as he did on other occasions, for she continued her late husband's interest in Port Arthur, Texas, civic affairs. Before her death in 1918, she had donated funds for the construction of a public library and a hospital and her will left additional funds to these institutions. (*Port Arthur, Texas, W.P.A. American Guide Series* [Houston: Anson Jones Press, 1940], pp. 80, 85, 87.)

26. Cullinan (Houston) to James L. Autry (Houston), November 20, 1913, Cullinan Papers.

27. Cullinan (Houston) to W. J. McKie (Corsicana), November 18, 1913, Cullinan Papers.

was about to sell them his own company's unissued stock and thus allow Standard Oil to seize control of the Texas Company. Lewis H. Lapham wrote Cullinan that the rumor "was widely circulated in Wall Street . . . and while we now have our differences, anyone associated with you, as I have these many years, knows the reports are completely false." Furthermore, Lapham hoped that "whatever the decision is in the proxy contest . . . you will remain with the company as its president."[28]

A few days before the meeting, it was clear that Cullinan faced defeat. Stockholders' proxies sent to him, mainly from Texas and other southwestern states, totaled only 42,738 shares.[29] On the other hand, John J. Mitchell, chairman of the eastern proxy committee, informed Cullinan that his group had proxies to vote over 220,000 shares.[30] This decisive rejection by the stockholders now left only one proper course of action open to Cullinan: resignation from the company. In his disappointment, he tried to be philosophical— even jovial. He wrote to an old Pennsylvania associate, Charles H. Todd, a few days before the meeting, in a vein reminiscent of his earlier petroleum career. "It was a good boarding-house brawl," he stated, "and some furniture was broken but our side was whipped fair and I'll be looking for another job soon."[31]

Although the result of the proxy contest was not in doubt, the stockholders' meeting held at the Texas Company offices in Houston on Tuesday, November 24, 1913, had its moments of corporate drama. The eastern contingent of directors and officers, led by John J. Mitchell, Arnold Schlaet, Elgood C. Lufkin, and James N. Hill, arrived by special railroad car two days previously. Cullinan presided as the meeting began at 11:00 A.M. The first item of business was the reading of a tribute to the late Charles G. Gates. The memorial praised "his most lovable character, his superior abilities . . . his democracy, generosity, honesty," and finally, "his busi-

28. Lewis H. Lapham (New York) to Cullinan (Houston), November 21, 1913, Cullinan Papers.

29. Proxy list dated November 22, 1913, Texas Company Folder, Autry Papers.

30. John J. Mitchell (Houston) to Cullinan (Houston), November 23, 1913 (telegram), Texas Company Folder, Autry Papers.

31. Cullinan (Houston) to Charles H. Todd (Washington, Pennsylvania), November 22, 1913, Cullinan Papers.

ness ability—innate and [as he was, after all, John W. Gates's son] inherited."[32] Cullinan then made a short statement noting that "since two-thirds of our stock is represented by the gentlemen making the second call for proxies" he was resigning as president and director of the Texas Company.[33] He thanked the company's stockholders and employees for their co-operation, and, according to a newspaper account, "waved all a 'pleasant good morning' and withdrew from the meeting."[34]

Elgood C. Lufkin then succeeded Cullinan as chairman of the meeting; whereupon James L. Autry made a short statement of resignation, both as a director and as the head of the company's Legal Department. Resolutions were then quickly passed which commended the services of Cullinan and Autry and offered them the "sincere thanks" of the stockholders. The meeting then proceeded with the election of directors. James N. Hill was elected to the vacancy left by the death of Charles G. Gates. John H. Lapham, brother of Lewis H., and Amos L. Beaty, the company's assistant general counsel, filled the vacancies left by Cullinan and Autry. The eastern majority also chose not to re-elect Texan Will C. Hogg. His place was taken by William A. Thompson, Jr., a company vice-president who headed the Marine Department.[35] At a directors' meeting later that day, Elgood C. Lufkin was elected president, succeeding Cullinan.[36]

32. Excerpts, Minutes of the Texas Company Stockholders' Meeting, November 24, 1913, Texaco Archives, XLVIII, 26.

33. *Ibid.*, 34.

34. Houston *Post,* November 25, 1913.

35. Excerpts, Minutes of the Texas Company Stockholders' Meeting, November 24, 1913, Texaco Archives, XLVIII, 35.

36. Excerpts, Minutes of the Texas Company Board of Directors' Meeting, November 24, 1913, Texaco Archives, XLVIII, 43.

Lufkin was born in Massachusetts in 1864; but since his father was an oilman, he grew up in the Titusville, Pennsylvania, area. He was graduated from the Massachusetts Institute of Technology in 1886, managed a Buffalo, New York, pump manufacturing concern, and joined the Texas Company in 1909 as manager of the Natural Gas Department. From 1910 through 1913, however, he had assisted Arnold Schlaet in the company's New York office. (Employee Biographies Folder, Texaco Archives.)

Because of Arnold Schlaet's "austere nature . . . he was not seriously considered as president of the company to replace Cullinan in 1913." (Unsigned and undated memorandum, Employee Biographies Folder, Texaco Archives.)

News of Cullinan's resignation spread quickly throughout the state. Two Houston newspapers chose to comment editorially and they lamented that his defeat was a further example of local domination by eastern economic interests. The *Post* thundered that an investigation should be undertaken by state authorities concerning the new affiliations of the Texas Company, "a corporation chartered by Texas—dealing with a natural source of wealth, developed by Texas citizens, financed in its beginning by Texas capital, and fructifying under the aegis of our Constitution and Laws. . . ."[37] The *Chronicle* also called for a state investigation of a company now seeking shelter in the "maelstrom of Manhattan," and it predicted that "New York water and Texas oil are not going to mix very well."[38]

But this initial bluster soon abated, and state authorities made no investigation. In the period immediately following their resignations, both Cullinan and Autry took their defeat philosophically and refused to inflame the issue further. Autry, still showing a tinge of sectional sensitivity, later wrote to state Representative Bryan T. Barry of Dallas that the "issue . . . was simply whether the company was to be operated from the executive standpoint from New York, according to New York ideas, or from Texas, according to Texas ideas. . . . " Autry admitted that "we made the issue ourselves, and were entirely prepared to abide the result, and to accept the situation cheerfully."[39] In reply to a letter of condolence from Governor Oscar B. Colquitt, Cullinan stated that "I was solely responsible for precipitating the issue that developed which . . . was brought about through looking from different viewpoints, i.e.—Texas and Eastern." He added that he would regret "any embarrassment to the new management that would follow as a result of our severing official connections with the company. . . ."[40]

But such regional commentary with its "Texas ideas" versus "New York ideas" obscures the major issue underlying the chain

37. Houston *Post*, November 25, 1913.

38. Houston *Chronicle*, November 26, 1913.

39. James L. Autry (Houston) to Bryan T. Barry (Dallas), December 3, 1913, Texas Company Folder, Autry Papers.

40. Cullinan (Houston) to Colquitt (Austin), November 29, 1913, Texas Company Folder, Autry Papers.

of events which forced Cullinan from the Texas Company. Cullinan, with his concept of highly personalized business leadership, honed in an earlier period, simply could not adjust to the group-decision-making demands of the company's Executive Committee. True, the growing importance of the New York office's administrative function, with its ready access to eastern investment capital, complicated Cullinan's task. But would Cullinan have been able to adjust to the sharing of executive responsibility within a growing and intricate corporate structure, even if that company had been wholly centered in Texas and solely owned by Texas investors? The answer must be in the negative—which is not to adjudge the validity of Cullinan's policies calling for unbridled expansion against those of the conservative-minded Executive Committee. The Texas Company, brought into being by Cullinan, was to experience even greater corporate expansion in its domestic and foreign operations within the next decade. Moreover, it was to do so under an executive group that Cullinan originally selected. His vigorous leadership had brought the company into corporate adolescence; his selection of men furnished a base for corporate maturity. Yet it cannot be doubted that the company's expansion would soon place even greater strains on Cullinan's style of individual leadership. In short, the Texas Company had outgrown one-man rule.[41]

It would be interesting to know more of Cullinan's plans for the Texas Company, had he been successful in the proxy battle and remained as the company's president. At the time of the proxy contest, he did not articulate a detailed design of future development and expansion. He had promised merely to continue the successful policies and practices which had brought the company

41. The efficiency of leadership in a corporation dominated by individual management as opposed to that of a concern where ultimate decision is made by a group or a committee still vexes contemporary students of our industrial order. For a recital of the evils which may befall a business under an "indispensable leader" and a defense of the virtues of group management, see Peter F. Drucker, *The Concept of the Corporation*, Mentor edition (New York: New American Library, 1964), pp. 35–42. Another recent writer, however, is highly critical of group management and feels that the committee system can become a major vice and a detriment to industrial expansion. See Clarence B. Randall, *The Folklore of Management*, Mentor edition (New York: New American Library, 1962), pp. 28–34.

growth in financial strength and resources.[42] It can be assumed, therefore, that, had Cullinan remained president, the patterns of development of the Texas Company would have remained basically unchanged. He would have striven to perfect a fully integrated petroleum company and to increase the sales of its diversified products in both domestic and foreign markets.

Yet, again, it should be emphasized that continued expansion inevitably would have brought further internal conflict if Cullinan had insisted upon retaining his highly personalized style of managerial direction. By the late nineteenth-century definition, Cullinan was an experienced oil executive with wide knowledge of the transportation, production, and manufacturing functions. But the rapid expansion of the American petroleum industry in the first decade of the twentieth century had enlarged this definition. The leader of a major oil company, for example, now needed intricate knowledge of an expanding domestic and foreign sales market. Cullinan had little experience of this type—indeed, few American oilmen outside of key Standard Oil executives then had such experience. He had already found Arnold Schlaet's advice on these problems invaluable, and further reliance on expert counsel would become a necessity in an increasingly complex and geographically dispersed organization. In time, Cullinan might have gracefully accepted the consequence that such advice, whether from individuals or from an Executive Committee, tends to supplant the decision of the single executive with the consensus of others. But his Texas career, launched at a time when few in that area could match his ability or experience in the oil industry, was based upon unchallenged personal leadership. Cullinan could not, or would not, accept these new consequences in making his decision to resign from the Texas Company in 1913.

When Cullinan left that company, he had no future plans. He took an extended trip to California where he visited his old Pennsylvania friend, Timothy M. (Tim) Spellacy, who had prospered in the development of that state's petroleum industry. "Spellacy

42. Statement to the stockholders of the Texas Company by J. S. Cullinan, president, November 17, 1913, Texaco Archives, XLVIII, 15.

tried to get me to stay," he confided to Judge McKie, "but my roots are in Texas now."[43] Cullinan was soon back in Houston and ready for work. An opportunity shortly arose which made it possible for him to re-enter the Gulf Coast oil industry. Ironically, this opportunity came as a result of his earlier disagreements with the Executive Committee of the Texas Company.

In 1912, the Texas Company had paid the partnership of Andrew D. Burt and Walter J. Griffith $50,000 for an undivided one-half interest in the 1,394 acres comprising the Jacob B. Stevenson tract near the Humble field, twenty miles north of Houston. This field was a salt-dome structure, and it had been rapidly exploited by the usual cycle of indiscriminate drilling and crowded well conditions at shallow depths about the top, or cap rock, of the dome. The Stevenson tract, however, was located three miles to the east of the cap rock and was untouched by the earlier exploitation. Cullinan was enthusiastic about this tract because he felt that deeper drilling along the flanks of the dome, as had already been shown at Spindletop, would yield valuable production. But the Executive Committee was cautious and conservative after the poor showing of the Texas Company in 1911 and 1912, and refused to appropriate funds for the tract's development. Burt and Griffith did not have enough capital to proceed on their own and tried several times to sell their remaining one-half interest to the Texas Company. Each offer, the last made in February 1914, was refused. Cullinan then stepped in. He purchased the undivided one-half interest held by Burt and Griffith for $10,000 in May 1914.[44]

That same month, Cullinan organized the Farmers Petroleum Company. Its authorized capital was $30,000. Cullinan held the largest percentage of Farmers stock, with a 33 1/3 percent interest; close associates Thomas P. Lee, Emerson H. Woodward,[45] Will C.

43. Cullinan (Bakersfield, California) to William J. McKie (Corsicana), March 2, 1914, Cullinan Papers. For further information on the activities of Tim Spellacy in the southern California petroleum development, see Gerald T. White, *Formative Years in the Far West: A History of Standard Oil Company of California and Predecessors Through 1919* (New York: Appleton-Century-Crofts, 1962), pp. 291–292.

44. Thomas J. Donoghue Memorandum, dated July 18, 1945, Texaco Archives, I, 19.

45. Both Lee and Woodward were former employees of the Producers Oil Company, the producing affiliate of the Texas Company, who resigned to join Cullinan in the

Hogg, and James L. Autry also held substantial interests. The charter specified that the new company was to be solely an exploration and producing concern. The interest acquired by Cullinan in the Stevenson tract at Humble was transferred to the company and plans were made for drilling five test wells later that year.[46] Cullinan also made arrangements with his erstwhile Texas Company associates concerning the exploration of the jointly held property. Obviously quite willing to allow Cullinan's company the cost of beginning development, the Texas Company further agreed to divide the tract into twenty-five-acre blocks. The blocks were then assigned to the two companies in alternating sections. Each company was to have exclusive rights to any production discovered within its assigned block.[47]

The five test wells drilled by Farmers Petroleum between August 1914 and January 1915 were productive at depths ranging from 3,000 to 3,200 feet. Four of the wells produced from 2,000 to 8,000 barrels daily and the fifth test well was a gusher reminiscent of the Spindletop boom. It initially flowed at a rate of 20,000 barrels a day.[48]

These successful tests by Farmers Petroleum stimulated a revitalization of the Humble field. In 1914, the field's output had slumped to 2,799,458 barrels. The next year, as other companies emulated the flank-drilling technique, the field's production rose to 11,061,802 barrels—the highest of all Gulf Coast fields—and was followed by 10,925,825 barrels in 1916.[49]

The Stevenson tract was one of the most prolific leases in this

---

formation of the Farmers Petroleum Company. Lee, the son of an oilman, was born at Petroleum, West Virginia, in 1871, and worked for Standard Oil companies in Ohio and Indiana before coming to Texas in 1903. Woodward, also the son of an oilman, was born at Podunk, New York, in 1879, and worked for Standard Oil in Pennsylvania and Ohio. He also came to Texas and went to work for Producers Oil in 1903. Both Lee and Woodward, experts in production methods, were to work very closely with Cullinan in the years subsequent to his resignation from the Texas Company. (*Oil and Gas Journal*, April 9, 1914, p. 28; January 28, 1915, p. 14.)

46. Charter and Incorporation Agreement, Farmers Petroleum Company, dated May 15, 1914, Farmers Petroleum Company Folder, Autry Papers.

47. *Oil and Gas Journal*, January 18, 1915, p. 14.

48. *Ibid.*; Houston *Post*, April 21, 1934.

49. Warner, *Texas Oil and Gas*, p. 372.

renewed Humble development. Farmers Petroleum obtained 831,-252 barrels of crude oil from this property in 1915, and 1,295,706 barrels in 1916.[50] Although this flush production depressed Gulf Coast crude oil prices to an average of forty-eight cents a barrel in 1915, the increased market demands of World War I shot oil prices upward. Gulf Coast crude oil prices thus increased to seventy-five cents a barrel in 1916 and to $1.07 a barrel in 1917.[51] Selling its output to the Texas and Gulf companies, whose pipelines linked Humble with their Port Arthur refineries, Farmers Petroleum made substantial profits during these years. The company paid $150,000 in cash dividends in 1914, and $225,000 during 1915. In addition, stock dividends had been paid in 1915, raising its capitalization to $100,000. During that year the company constructed a lease gathering system at Humble and completed a fifteen-mile pipeline to an extensive tank farm (forty-eight 55,000-barrel steel tanks) at East Houston. Property had been purchased along the Houston Ship Channel for tidewater loading facilities and a pipeline spur from the East Houston tank farm to the channel was under construction.[52]

The success of Farmers Petroleum soon led to its corporate demise. As Cullinan had found with his first Spindletop venture, the Texas Fuel Company, the capitalization and corporate powers of Farmers Petroleum were too limited for further development and business expansion. On March 8, 1916, two new companies were organized: the Republic Production Company was chartered under Texas law as an exploration and production company with authorized capital of $3,000,000; and the American Petroleum Company was chartered under Texas law as a pipeline and refining company with authorized capital of $3,000,000. The Farmers Petroleum Company was then dissolved and its assets sold to the new companies. Republic Production purchased, for $1,500,000 of its stock, the producing wells and lease rights formerly held by Farmers Petroleum; American Petroleum purchased, also for $1,-

50. Production Records, Farmers Petroleum Company Folder, Autry Papers.

51. Williamson, et al., The American Petroleum Industry, 1899–1959, p. 39.

52. Financial Statements, 1914, 1915, Farmers Petroleum Company Folder, Autry Papers.

500,000 of its stock, the pipeline and storage units originally constructed by Farmers Petroleum.[53]

The chain of transactions flowing from the dissolution of Farmers Petroleum was not yet completed, however. On May 19, 1916, the American Republics Corporation was organized. This concern was chartered under Delaware law with a total authorized capital of $10,000,000: $5,000,000 in common stock and $5,000,000 in preferred stock. The new corporation was a holding company; its function was "to hold the stocks of subsidiary companies transferred, sold, or assigned to it."[54] The trustees of the dissolved Farmers Petroleum Company then sold the Republic Production and American Petroleum stock for $3,000,000 of the common stock of the holding company. This American Republics stock was then distributed to the former stockholders of Farmers Petroleum in proportion to their stock interest in that dissolved company.[55]

Although it was not an uncommon form of business organization, there were several reasons why Cullinan and his legal advisers chose to incorporate a holding company chartered under out-of-state law. First and foremost, attorney James L. Autry felt the Delaware charter was necessary "to avoid the uncertainties of local law which prohibit" the incorporation of a holding company, and thus he advised that "we should use foreign law to charter a corporation not specifically authorized by Texas statutes."[56] Sec-

53. Application of American Republics Corporation to Committee on Stock List, New York Stock Exchange, November 5, 1923, Cullinan Papers.

54. Charter, American Republics Corporation, dated May 19, 1916, Cullinan Papers.

55. The original $5,000,000 in preferred stock authorized by the charter of American Republics Corporation was issued between July 1916 and December 1919. It was used either to purchase stock of subsidiary corporations or sold to raise funds for general corporate purposes. The $2,000,000 worth of common stock which remained, after the purchase of the Republic Production and American Petroleum properties, was retained as unissued treasury stock and distributed as a stock dividend in June 1918. (Application of American Republics Corporation to Committee on Stock List, New York Stock Exchange, November 5, 1923, Cullinan Papers.)

56. Autry (Houston) to Cullinan (Houston), March 18, 1916, Cullinan Papers. In this letter, Autry also pointed out that the advantages of a holding company as to centralized control of several subsidiary companies could be obtained through the formation of an "unincorporated trusteeship," but this type of organization "would have few, if any, of the legal advantages of incorporation." An examination of Texas in effect bears out Autry's contention that there was no specific

ond, close managerial control of subsidiary corporations was obviously possible under a holding company, particularly as Texas authorities still insisted upon a corporate separation of producing operations from transportation and manufacturing functions. Cullinan, as president and as the largest stockholder of the American Republics Corporation, could thus effectively control its subsidiary organizations. And third—a new factor which was then just beginning to be recognized in American business—the holding-company structure possessed advantages under federal income tax laws by allowing, as Autry noted, "for the consolidation of losses from unprofitable operations with the income of profitable subsidiaries and . . . result in a reduced tax burden."[57]

Events soon made Cullinan well aware that federal income tax laws were now, indeed, a potent factor in business dealings. Upon the formation of the American Republics Corporation, the Texas oilman had received stock in that company worth, at par value, $1,598,400 for an interest in the Farmers Petroleum Company valued at only $26,640. Federal tax officials claimed that the difference between these two amounts represented taxable income and assessed him an additional $156,216.66 in personal income tax for 1916. Cullinan paid the tax and challenged the assessment. Although he received an adverse judgment upon trial in federal district court, the decision was ultimately appealed to the United States Supreme Court. Finally on April 30, 1923, that court affirmed the lower court's decision and held that the exchange of stock between Farmers Petroleum and American Republics (via the Republic Production and American Petroleum companies) was "a sale involving taxable gain . . . and that Cullinan, in a legal sense . . . became taxable on it as income for the year 1916."[58]

In addition to the Republic Production and American Petroleum companies, American Republics acquired three other wholly

statutory authorization for the incorporation of a holding company. See John G. McKay, *Principal Corporation Laws of the State to Date* (Austin: A. C. Baldwin & Sons, 1916), pp. 3–10.

57. Autry (Houston) to Cullinan (Houston), March 18, 1916, Cullinan Papers.

58. *Cullinan* v. *Walker, as Collector of Internal Revenue for the First District of Texas*, 262 U.S. 134 (1923).

owned subsidiaries in 1916: the Papoose Oil Company, organized to operate oil properties in Oklahoma; the Federal Petroleum Company, formed to operate Louisiana oil properties; and the Petroleum Iron Works Company, Cullinan's old Washington, Pennsylvania, oil equipment manufacturing concern. This latter acquisition was to prove increasingly valuable as, through several of its own subsidiaries, manufacturing functions ultimately were broadened to include the building of railroad tank cars, freight cars, steel shipping containers, and the eventual operation of a shipyard in Beaumont, Texas. "American Republics," Cullinan confided to Autry, "is off to a good start . . . and will give us a firm base for all types of petroleum operations."[59] He was also particularly pleased because the enterprise was formed without "asking for one penny of Eastern capital."[60] Profits from the Farmers Petroleum Company and funds received from periodic sales of Texas Company stock were undoubtedly sufficient to launch American Republics.[61]

But by far the most significant development of American Republics' first year was an operating agreement concluded on November 15, 1916, between its producing subsidiary, Republic Production Company, and the Houston Oil Company. This agreement gave Cullinan's company an undivided one-half interest in almost 800,-

59. Cullinan (Houston) to James L. Autry (Houston), January 17, 1917, Cullinan Papers.

60. Cullinan (Houston) to Autry (Houston), October 31, 1916, Cullinan Papers.

61. As noted previously, Cullinan held 7,134 shares of Texas Company stock when he resigned in 1913. His first reaction at that time was to make the severance from the company complete by selling all his stock at once. Yet, upon reflection, he decided it would be wiser to sell his holdings gradually as he needed funds for other investments. Also, the sudden "dumping" of his large holdings might have weakened, temporarily, at least, the market value of the Texas Company shares. While Cullinan was bitter about the circumstances which forced him to leave the company, he had no desire to cause the Texas Company any further embarrassment. Thus he disposed of his shares gradually, particularly during 1914 and 1915, as he needed funds for his Farmers Petroleum and American Republics ventures. At the end of 1915, he held 2,993 shares. By 1920, his holdings were reduced to 800 shares. (Interview with James H. Durbin, New York City, August 23, 1962; List of Texas Company Stockholders, November 15, 1913; December 31, 1915, Texas Company Folder, Autry Papers; Federal Income Tax Return [copy] of J. S. Cullinan for Calendar Year 1920, Cullinan Papers.)

ooo acres of east Texas timberlands with exclusive rights to explore and develop these lands for petroleum purposes.[62]

The Houston Oil Company of Texas, capitalized at $30,000,000, was chartered in July 1901, at the height of the Spindletop boom. Its creation resulted from the ideas of John Henry Kirby, a colorful, personable young Texas attorney and lumberman. On borrowed eastern capital, Kirby had accumulated substantial holdings of virgin pine lands in the fourteen east Texas counties immediately north of the Beaumont-Spindletop area. These lands, consisting of over 800,000 acres, were subsequently transferred to the Houston Oil Company. Kirby organized another corporation, the Kirby Lumber Company, which was a lumbering firm with a dozen sawmill plants scattered throughout the east Texas "piney woods" area. A long-term timber contract was then executed between the two companies. The Kirby Lumber Company was to purchase and cut timber on the lands of the Houston Oil Company. The proceeds from the sale of this timber would give Houston Oil a guaranteed income and, in turn, attract further capital for petroleum exploration and development upon its lands.

But even to get these plans under way required substantial capital—far more than Kirby could raise with his funds mainly tied up in timberland. Thus Kirby enlisted the help of a co-promoter, Patrick Calhoun, a grandson of South Carolina's John C. Calhoun. Patrick Calhoun had long since left his native South for a distinguished career as a New York attorney specializing in railroad and street railway consolidations. He had access to eastern capital and promised to raise enough money for Kirby's plans in return for a half interest in the enterprise.

The union of these two native but dissimilar sons of the South—one, the scion of antebellum Carolina aristocracy, now a Wall Street attorney, the other a hearty, self-made Texan from the "piney woods" country—was not long to endure, however. The end of the Spindletop boom made eastern banking interests suspicious of oil promotions in unproven areas, however close they might be

62. The formation and history of the Houston Oil Company prior to the 1916 operating agreement is summarized from the author's *The Early History of the Houston Oil Company of Texas, 1901–1908*, pp. 5–88.

to the Spindletop development. As a result, Calhoun failed to find sufficient capital for Houston Oil to develop its properties. Kirby, meanwhile, was also handicapped by lack of operating capital. His lumbering concern, the Kirby Lumber Company, could not meet the terms of the timber contract with the oil company. By 1904, both companies were in receivership and there followed years of litigation—suit and countersuit—employing a corps of attorneys and masters-in-equity in attempts to straighten out the tangled affairs of both companies.

A settlement was finally reached in 1908. The financial affairs and management of the two companies were separated: Kirby was placed in full control of Kirby Lumber and Calhoun was given a substantial interest in the Houston Oil Company. The timber contract was modified under terms that made its execution profitable for both companies; and during the next years, Houston Oil made substantial progress toward reducing its indebtedness. Calhoun, meanwhile, disposed of his interest in the company. Houston Oil was then controlled by a Baltimore bank, through bonds held by the Maryland Trust Company, and a group of St. Louis investors who had been persuaded to trade their Corsicana oil properties (the Southern Oil Company) for stock in Calhoun's enterprise.[63] Although these Corsicana properties gave Houston Oil minor petroleum income, the company was hardly worthy of its name in 1916. Instead of being an oil company, it was more a vast and heavily mortgaged "land company" of 800,000 acres, whose major source of income was from the timber contract with the Kirby Lumber Company. Its managers at its home office in Houston, Texas, were, therefore, not oilmen but, primarily, lumbermen, their major task being to execute and supervise the timber contract with the Kirby Lumber Company.

Joseph S. Cullinan had watched closely the fortunes of the Houston Oil Company during these years. In 1902, he had offered to undertake petroleum development on the company's lands in exchange for Calhoun's purchase of Texas Company stock. But Calhoun had problems enough trying to obtain capital for his ventures

63. For the role of the Southern Oil Company in the development of the Corsicana field, see Chapter 3, pp. 62–66.

with Kirby, and the offer was refused.[64] A few years later, in 1907, Cullinan again approached Houston Oil officials concerning the development of their property. But Thomas H. Franklin, the company president, replied that "while your offer is interesting . . . little can be done until the present litigation [between Kirby and Calhoun] is concluded."[65]

In 1913, following his resignation from the Texas Company, Cullinan renewed his overtures to the Houston Oil Company. He wrote the company's attorney that he would bear the "entire cost of petroleum development" on the company's lands and that such an agreement "could not help but have a beneficial effect on all parties."[66] Still, the local officials of Houston Oil, who were mainly lumbermen fathoming little about the mysteries of the oil business, were understandably reluctant to rush into an agreement. Cullinan played a waiting game. He particularly courted Houston Oil's general manager, Amos W. Standing, an elderly lumberman who was more than a little flattered by the attention he received from the recognized "dean" of the Texas petroleum industry.[67] Cullinan's patience and perseverance eventually paid off: an operating agreement between Houston Oil and Cullinan's Republic Production Company was finally concluded in November 1916.

This agreement gave Republic Production exclusive rights to petroleum development on lands owned by the Houston Oil Company. Cullinan's company was to bear the entire cost of this development; and the company also specifically pledged to spend a minimum of $250,000 for this purpose during the succeeding four-year period. Houston Oil received an undivided half interest in any petroleum discovered, and, in turn, it conveyed to Republic Production an undivided half interest in the entire 800,000 acres owned by the oil company.[68]

64. Joseph P. Dobbins (New York) to Cullinan (Beaumont), May 1, 1902, Cullinan Papers. Dobbins was a law partner of Patrick Calhoun.

65. Thomas H. Franklin (Houston) to Cullinan (Beaumont), April 27, 1907, Cullinan Papers.

66. Cullinan (Houston) to Lloyd E. Denman (San Antonio), December 30, 1913, Cullinan Papers.

67. Interview with Charles A. Warner, November 17, 1963, Houston, Texas. Mr. Warner was a former officer of Houston Oil who joined that company in 1919.

68. Operating Contract and Conveyance between the Houston Oil Company of

Tests were immediately begun on the Houston Oil Company lands, and in July 1918, Republic Production drilled the discovery well which opened the prolific Hull field in Liberty County, Texas. On one tract of the Houston Oil Company lands at Hull, the famous Dolbear Lease of 800 acres, Republic Production produced a total of 4,695,415 barrels of crude oil from 1918 through 1923.[69] Further Republic Production development of the Houston Oil lands throughout the 1920s resulted in the discovery of the productive Arriola, Spurger, and Silsbee fields in Hardin and Tyler counties. The agreement further specified that after the initial drilling was completed by Republic Production, either party could select units of the property to develop as the operator. The other party could either join the venture by contributing one half of the development expenses or decline to pay the expenses but receive a royalty on resulting production from such tracts. The Houston Oil acreage continued to yield valuable production through ensuing decades. The Joe's Lake and East Village fields in Hardin County were two of the most important discoveries in the Gulf Coast area of Texas after the Second World War.[70]

The operating agreement between Houston Oil and Republic Production was thus a boon to both companies. Although Houston Oil still derived income from its timber contract with the Kirby Lumber Company, the agreement with Republic Production gave the company a high degree of diversification, and it soon branched into further oil and natural gas ventures in central and south Texas.[71] The advantages of the agreement to Cullinan's company were obvious. By obtaining exclusive development rights and a half interest in 800,000 acres of land, Cullinan was relieved of the financial burden of expending large sums in annual lease rentals necessitating costly—and perhaps hasty and inconclusive—exploration procedures. The fact that these properties later became very pro-

Texas and the Republic Production Company, dated November 15, 1916, HOCO, Warner Collection.

69. Application of American Republics Corporation to Committee on Stock List, New York Stock Exchange, November 5, 1923, Cullinan Papers.

70. Warner, *Texas Oil and Gas*, pp. 213–214; James H. Durbin (New York) to author, August 23, 1966.

71. King, *The Early History of the Houston Oil Company, 1901–1908*, p. 89.

ductive reinforces the conclusion, as the late Charles A. Warner colorfully put it, "that Cullinan's deal with Houston Oil was one of the great 'horse trades' of Texas petroleum history."[72]

A similar pattern followed in 1921, as an American Republics subsidiary, Federal Petroleum Company, made an agreement with the Frost Lumber interests of Shreveport, Louisiana. For undertaking the expense of development, Cullinan's company received a half interest in the mineral rights to over 230,000 acres of timberlands in northwestern Louisiana. Subsequent exploration did not yield substantial oil production; but in 1924, a large natural gas reservoir was discovered on Frost lands near Monroe in Union and Ouachita Parishes, Louisiana. Cullinan planned to market the gas through construction of a pipeline to New Orleans. Trying to find financial support for this project, he organized with Frost a joint venture, the Union Power Company, at Monroe, to manufacture carbon black, then the most effective method to utilize immediately natural gas without transmission facilities. But in 1925, as the pipeline project bogged down in financing problems, Cullinan and Frost decided to sell their natural gas interests in the Monroe field for $2,500,000 to Interstate Natural Gas Company, a subsidiary of the Standard Oil Company (N.J.). Cullinan had made a nice profit, but eventually Interstate did complete the gas transmission line to the lucrative New Orleans market that he had initially planned.[73]

Ironically, these successes seemed to stunt Cullinan's initial plans involving American Republics and its subsidiary, the American Petroleum Company. At that time, he talked of planning integrated petroleum operations involving "transportation, refining, and product marketing . . . similar to those we fashioned in the Texas Company."[74] Cullinan made no prolonged effort to do this, however.

In 1919, he acquired the controlling interest in the Galena Signal Oil Company of Franklin, Pennsylvania. This company was an old Standard Oil subsidiary which once enjoyed a virtual mo-

72. Interview with Charles A. Warner, November 17, 1963, Houston, Texas.

73. Annual statements, 1921–1925, American Republics Corporation, Cullinan Papers; James H. Durbin (New York) to author, August 23, 1966.

74. Cullinan (Houston) to James L. Autry (Houston), June 17, 1916, Cullinan Papers.

nopoly in the manufacture of railroad lubricants. Cullinan soon organized a subsidiary of American Republics, Galena Signal Oil Company of Texas, to operate a small refinery along the Houston Ship Channel. Yet, little of a sustained attempt was made to expand or diversify these manufacturing operations. Galena Signal Oil of Pennsylvania was sold subsequently to the Valvoline Oil Company. Cullinan continued to operate the Houston-based Galena Signal Oil Company of Texas for a few years, but eventually it was sold to the Texas Company in 1928.[75]

To have expanded the scope of the operations within American Republics and its subsidiary companies would, of course, have posed the problem of raising additional capital—a problem, however, that would not seem insurmountable in view of Cullinan's contacts. The most probable reason for his refusal to attempt further expansion was that he was satisfied with things as they were. As president of the Texas Company, he had been uncomfortable heading an organization whose major problems were no longer simply "finding" oil but involved as well the intricate patterns of marketing and administration. The American Republics Corporation and its subsidiaries were dominated by local capital and Cullinan's unchallenged leadership. Moreover, particularly through the agreement with Houston Oil, the exploration and producing functions were to become paramount. Cullinan's new organization was, in short, an oilman's enterprise, and he preferred that its major function remain the exciting quest for new petroleum discoveries.

Thus, in the formation of the American Republics Corporation, Cullinan began his last significant venture in the southwestern petroleum industry. Although the Petroleum Iron Works Company in Pennsylvania contributed some income, by 1923 the operations of American Republics centered about three producing subsidiaries: the Republic Production Company, which not only operated on the Houston Oil Company lands but had acquired lease rights to an additional 220,000 acres within Texas; the Federal Petroleum Company, which held lease rights to 266,853 acres in Louisiana; and the Papoose Oil Company, which leased 5,388

75. Houston *Post*, April 21, 1934; James H. Durbin (New York) to author, October 21, 1969.

acres in Oklahoma. The combined crude oil production of these properties from 1918 through 1922 totaled 12,240,975 barrels.[76]

Yet even as Cullinan concentrated upon petroleum exploration and development activities in the southwest, through American Republics he still controlled a group of northern companies consisting of his old oil equipment manufacturing concern, Petroleum Iron Works of Pennsylvania, and its subsidiaries. Although one of these subsidiaries, Pennsylvania Tank Line, had made substantial profits through railroad tank car construction and leasing, the northern companies, as a group, had been only moderately profitable for many years. Cullinan was well aware of this and had planned eventually to consolidate the Pennsylvania business by eliminating its unprofitable operations.

In 1927, a group of American Republics' Texas stockholders, led by long-time associates Thomas P. Lee, Emerson H. Woodward, and Will C. Hogg, sought to force this issue. Claiming that the northern companies were losing substantial money at the expense of southwestern oil operations, these dissidents demanded immediate sale of the Pennsylvania properties and were prepared to oust Cullinan as president of American Republics if he opposed this action.

Stung by this challenge from old friends, Cullinan characteristically resisted this attempt to dictate managerial policy, and an acrimonious proxy battle soon raged for control of American Republics. The dissenting group particularly sought to influence other stockholders by hurling charges of nepotism at Cullinan. A year before, he had made his son, Craig F. Cullinan, vice-president and director of American Republics, while a younger brother, John Francis (Frank) Cullinan, then served as president of the Republic Production Company. In an open letter to the stockholders, Cullinan struck back by pointing out that, after graduation from Yale in 1917 and military service during World War I, Craig Cullinan had compiled a record as an independent manager of subsidiary operations which well merited his promotion. Furthermore, Frank Cullinan had more than thirty years' experience in the petroleum

76. Financial Statement, American Republics Corporation, dated September 30, 1923, Cullinan Papers.

industry, and he had also been eminently successful in directing independent producing ventures.[77]

This vigorous defense carried the day. By stockholders' vote of March 1927, a Cullinan slate of directors was elected; Cullinan was subsequently re-elected president, and the dissidents were defeated. But Cullinan never again resumed business relations with Lee, Woodward, or Hogg, all of whom, within a short time after the fight, disposed of their stock holdings in American Republics Corporation.[78]

In 1929, however, Cullinan voluntarily relinquished the presidency of American Republics to his son, Craig. Later, as the company was released from receivership during the depression years, Cullinan was re-elected president, and he served in that capacity until 1936, when his son again succeeded him. Craig F. Cullinan was to serve as president of American Republics until his death in 1950. He was, in turn, succeeded as president by Torkild Rieber, who served until 1956, when American Republics and its subsidiaries were liquidated and its assets sold to Sinclair Oil Corporation for $108,000,000.[79]

Throughout the later stages of his career, Joseph Cullinan continued his fight for the conservation of oil and gas through the elimination of wasteful drilling and production practices. In 1916, new fields in north Texas were being rapidly exploited through the usual cycle of indiscriminate drilling; and yet, against the background of World War I, petroleum had assumed new importance as a military and naval fuel. In exasperation, Cullinan wrote a polemic calling for federal control of the nation's oil industry. He argued that Congress had the unquestioned power to exercise the right of eminent domain over all the country's petroleum deposits

77. Letter to the Stockholders of American Republics Corporation from J. S. Cullinan, February 24, 1927, Cullinan Papers. These charges of nepotism from the dissident group were very irksome to Cullinan, as Thomas P. Lee had employed two of his brothers and a nephew in executive positions with American Republics subsidiaries; Emerson H. Woodward's brother-in-law was superintendent of production for the Republic Production Company in the Hull field of east Texas. (James H. Durbin [New York] to author, August 23, 1966.)

78. James H. Durbin (New York) to author, August 23, 1966.

79. *Ibid.*; Annual reports, 1936, 1956, American Republics Corporation, Warner Collection.

and that it had also the authority to condemn and acquire all oil pipelines and other forms of transportation utilized by the industry. Furthermore, federal authority was sufficient to legislate a "reasonable price between consumers, royalty owners, producers, transporters, manufacturers, and distributors for domestic consumption."[80]

However, this open demand for the virtual nationalization of the country's petroleum industry created but a slight stir. Of the major Texas newspapers, only the Dallas *Morning News* chose to reply editorially to Cullinan. That newspaper chided him as being too pessimistic about the future discovery of new oil fields which would insure a sufficient national supply of petroleum. Moreover, should the supply of oil falter, new forms of energy such as "electricity, water power and the harnessing of tides will replace oil."[81]

As the Texas Railroad Commission assumed further control over well-spacing, drilling practices, and, finally, in the early 1930s, the allocation of production, Cullinan modified his position that only the federal government could bring order to the petroleum industry. "State agencies that are staffed by expert personnel," he declared, "are probably better qualified to regulate the industry." Yet, even then, he maintained, there must be "constant vigilance to insure that politics does not interfere."[82] Cullinan's belief in the need for governmental regulation of the petroleum industry shifted from emphasis on federal control to state control, as a result of his growing concern with national political trends of the early 1930s. Cullinan took pride in his lifelong membership in the Democratic party; he had been a county chairman in Pennsylvania before coming to Corsicana and he had continued this orthodoxy throughout his Texas career—with one major exception. After the First World War, the resurgence of the Ku Klux Klan, with its anti-Catholic, anti-foreign-born bias, greatly disturbed Cullinan. In 1922, he refused to support the Democratic candidate for the United States Senate, Earle B. Mayfield, who was supported by the Texas Klan. Cullinan crossed party lines and became a major financial contribu-

80. *Oil and Gas Journal*, May 25, 1916, p. 31. This article first appeared in the Houston *Chronicle*, April 10, 1916.
81. Dallas *Morning News*, April 18, 1916.
82. Cullinan (Houston) to John S. Allyn (St. Louis, Missouri), November 10, 1933, Cullinan Papers.

tor to the spirited but unsuccessful campaign of the Republican candidate, George E. B. Peddy, of Houston.[83]

During these months, even the sentimental Irish-American observance of St. Patrick's Day was an occasion to defy the Klan. On March 17, the Irish national flag always flew at Shadyside, the Cullinan home at South Main Street and Montrose Boulevard in Houston. But in 1922, Cullinan began the yearly tradition of flying the black skull-and-crossbones flag atop the Petroleum Building, his headquarters in downtown Houston. When amused and puzzled citizens periodically asked him for explanation, Cullinan always made the same terse statement: "The display of the Jolly Roger is intended," he said, "as a warning to privilege and oppression within or without the law—the latter including witch-burners, fanatics, and the like who fail to realize or ignore the fact that liberty is a right and not a privilege."[84]

But the eventual eclipse of Klan political power in Texas brought Cullinan back into the Democratic fold. He avidly supported the Democratic presidential candidacies of Alfred E. Smith in 1928 and of Franklin D. Roosevelt in 1932, despite a close friendship with Republican Herbert C. Hoover which stemmed from Cullinan's service as an advisor to the Food Administration during the First World War. However, his support of Roosevelt was of short duration. Cullinan soon turned against the New Deal and charged that its early programs tended in the direction of Socialism by aiding the spread of centralized and invisible government.[85] In 1935, he became a charter member of the anti-New Deal American Liberty League; and in the presidential campaign of 1936, he was a major financial backer for the "Texas Democrats for Landon." According to a close associate, Cullinan did not explain these defections as indicating any changes in his own political position. It was the programs of the Roosevelt administration which were "incompatible with his views as a businessman, as an individualist, and a lifelong Democrat."[86]

83. Houston *Press*, October 10, 1922.
84. Houston *Post*, March 11, 1937.
85. Cullinan (Houston) to William B. Lawson (New York), August 10, 1934, Cullinan Papers.
86. James H. Durbin (New York) to the author, July 8, 1961.

In addition to his interest in political affairs, Cullinan was active in business, civic, and cultural organizations. He had prophesied as early as 1905 that Houston would become the major petroleum center of the Southwest, and he well realized, from his Spindletop days, the stimulation that low-cost water transportation gave industrial development. During his tenure as president of the Houston Chamber of Commerce from 1913 through 1919, he was thus particularly energetic in his zeal to obtain federal appropriations for the construction and maintenance of the Houston Ship Channel. He frequently visited Washington during these years to give testimony before various congressional committees and other governmental bodies, particularly the staff of the United States Corps of Engineers. His efforts were successful; through federal appropriations matched with funds voted by local bond elections, the ship channel was deepened and, in 1915, Houston, more than fifty miles inland from the Gulf of Mexico, became a salt-water port. In 1922, Cullinan further spurred the development of the ship channel area by building the Houston North Side Belt Railway, a short switching-and-service line designed to connect the north shore of the channel with existing trunk-line facilities. This keen interest in water transportation for the Southwestern Gulf area also led him to take an active role during the 1920s and 1930s as an officer and director of the Intracoastal Canal Association of Louisiana and Texas and of the National Rivers and Harbors Congress.[87]

As a civic leader and a man of wealth,[88] Cullinan supported a

87. Houston *Chronicle*, March 12, 1937.

88. Although it is obviously difficult to compare Cullinan with contemporary "Texas millionaires" because of the involved problems of transferring values of the 1920s and 1930s to present standards, he was unquestionably a wealthy man for his time. In 1913, his gross income for federal income tax purposes was $173,968, including $30,000 annual salary paid him as president of the Texas Company. During the succeeding seventeen years, from 1914 through 1930, his gross income for tax purposes, excluding taxable gains from sales of property, averaged $278,412 annually, which was mainly from dividends paid by his American Republics Corporation and its subsidiaries. The depression years brought a substantial decrease—he reported a taxable income of only $20,119 in 1932—but his taxable earnings rebounded to an annual average of $170,000 from 1934 through 1936. The value of his estate for inheritance tax purposes at his death in March 1937 was $4,320,640. This included approximately $3,500,000 in securities, of which $2,900,000 was in the stock of American Republics Corporation. The residue of his estate was mainly in real estate.

great number of cultural and philanthropic activities. He was particularly generous in contributions to the Houston Art Museum, the Houston Symphony Orchestra, and the Houston Negro Hospital, which he endowed as a memorial to his son, John Halm Cullinan. This son had served as a field artillery officer in the American Expeditionary Forces in France in 1918 and died in Sharon, Pennsylvania, of a pulmonary ailment in 1920.[89] Cullinan was appointed chairman of the Mount Rushmore National Memorial Committee in 1928 and served in that capacity until 1933. During the First World War, the Texas oilman served as a special advisor to the Food Administration under Herbert C. Hoover.[90]

On a trip to California in the spring of 1937, Cullinan stopped in San Francisco, intending to visit his wartime associate in the Food Administration, former President Hoover. While he was staying in a hotel, the San Francisco area experienced a sharp earth tremor during the night of March 7, 1937. The hotel guests fled into the night air, and Cullinan caught cold. The cold soon deepened into pneumonia; and J. S. Cullinan, in his seventy-seventh year, died in a Palo Alto, California, hospital on March 11, 1937.[91]

The rapid development of the Texas petroleum industry from the late 1890s through the early Twentieth century offered numerous opportunities to experienced managerial talent migrating into the state. Joseph S. Cullinan, a Pennsylvanian, well-schooled in petroleum operations by a long apprenticeship in eastern oil fields, was one of the first oilmen to accept this challenge. The problems faced by Cullinan were thus the problems of a young industry, literally explosive in the pace of its growth and achievement.

(Federal income tax file; certified copy of Estate Inventory, March 11, 1937, Cullinan Papers.)

89. The largest charitable bequest in his will was also a substantial contribution ($524,000) to the Houston Negro Hospital. (Certified copy of Will of J. S. Cullinan, Cullinan Papers.)

90. Houston *Chronicle,* March 12, 1937.

91. His immediate survivors were a son, Craig F. Cullinan; three daughters, Miss Nina J. Cullinan, Mrs. Andrew Jackson Wray (Margaret), and Mrs. Rorick J. Cravens (Mary). His wife, Lucy Halm Cullinan, died in 1929 and his eldest son, John Halm Cullinan, as noted, died in 1920. (Houston *Post,* March 15, 1937.)

Cullinan, an active and aggressive personality, was successful in the accommodation of influential local interests, in the recruitment of eastern capital when Texas capital proved insufficient, and in his quest for industrial stability through the consolidation and integration of the various phases of petroleum operations. His concern about the exploitative nature of the early Texas oil field development led to legislation which began the state's first petroleum conservation laws. Yet there were disappointments and failures. He failed to gain legislative approval legitimatizing further integration and consolidation within the petroleum industry, and he left the presidency of the thriving oil company he had founded because his highly individualized style of managerial direction could not adjust to the demands of a growing and complex organization. Thus, a study of Cullinan's career, in its triumphs and failures, well reflects the patterns and pressures resulting in change and growth within the Texas petroleum industry during these crucial years of its development.

Bibliography

Manuscript Collections

Several important collections were used in the preparation of this work. Among them were the Joseph Stephen Cullinan Papers, now in the Texas Gulf Coast Historical Association Archives at the University of Houston. These papers, which include both personal correspondence and business records, were particularly valuable in tracing the organization and development of the Texas Company during Cullinan's presidency from 1902 to 1913. Records in this collection of other phases of his career are somewhat limited. There is little on his earlier activity in the Pennsylvania and Ohio oil fields. Contemporary records of his first Texas petroleum ventures at Corsicana from 1897 to 1901 are sparse, although important information concerning that development can be gleaned by close examination of subsequent correspondence between Cullinan, after he had moved on to the Spindletop area in 1902, and former Corsicana associates. These papers are essential in revealing the dynamic mood and pace of the Texas petroleum industry in the early twentieth century.

Complementing the Cullinan Papers are the James Lockhart Autry Papers at William Marsh Rice University. Autry served as legal counsel to Cullinan from 1897 to 1920; thus, his papers reveal the problems involved in adapting Cullinan's ambitious plans to prevailing incorporation and antitrust statutes.

The Texas Company Archives, located in the company headquarters in New York, contain extensive material covering Cullinan's tenure as the company's president. Most useful were the business correspondence, excerpts of corporate minutes, financial statements, production and manufacturing statistics, and memoranda prepared by officers and employees covering company history from 1902 through 1913. Because of the sheer bulk and scope of these records, I was selective in the documents examined, attempting to concentrate primarily on Cullinan's activities, not on writing an extensive institutional history of the company during this period. To date, no scholarly or objective study has been made of the Texas Company. The extent of the material in the company's archives, however, indicates that such a study would make a significant contribution to American business history.

The Charles A. Warner Collection, held, like the Cullinan papers, by the Texas Gulf Coast Historical Association Archives, contains much of value to researchers in petroleum history. Notes and documents used by this prominent oil historian—author of *Texas Oil and Gas Since 1543*—and the records of the Southern Oil and Houston Oil companies deal with many facets of Cullinan's activities and were used extensively in this work.

The papers of the late Louis Wiltz Kemp at the University of Texas were also examined. In addition to being an inveterate collector of Texas data, Kemp was a long-time employee of the Texas Company and took particular interest in its early history. Most of his material in this collection pertaining to petroleum history, however, is duplicated from the Texas Company Archives, as Kemp was the prime organizer of that collection.

Interviews and Letters

Interviews with petroleum authorities, members of the J. S. Cullinan family, and Cullinan's former business associates were particularly rewarding and helpful. Two interviews with James A. Clark—co-author with Michel Halbouty of *Spindletop*—in Houston, Texas, on November 5 and December 10, 1964, obtained further information concerning Cullinan's role in the early Spindletop development. Various interviews with a daughter of J. S. Cullinan, Nina J. Cullinan, were essential to an understanding of her father's family background and history. A conference in New York City on August 23, 1962, with James H. Durbin, a former officer of the American Republics Corporation, brought further insight into Cullinan's personality and business practices. Information concerning corporate accounting procedures prior to the advent of federal income tax laws in 1913 was obtained from J. Robert Mann, C.P.A., at Houston on July 13, 1964. An interview on October 15, 1963, with Carlton D. Speed, Jr., petroleum consultant and geologist, was very helpful on the Corsicana, Texas, oil field development. And, finally, three conferences with Charles A. Warner at Houston on January 16 and November 17, 1963, and June 10, 1964, were very important in giving further enlightenment on such subjects as the geologic structure of Texas and the consequences of the state's early petroleum conservation laws.

In addition, I have in my possession two important letters with information on phases of Cullinan's career. One, from James H. Durbin, New York City, July 8, 1961, dealt with Cullinan's family history, his educational background, and his pre-Texas career in the Pennsylvania oil fields. The other letter, dated June 4, 1963, from Sharon, Pennsylvania, was from James A. Connelly, an old friend and business associate of Cullinan in the Petroleum Iron Works. This letter also gave important details of Cullinan's early career in Pennsylvania.

Unpublished Manuscripts

One of the most significant of the unpublished manuscripts used in this work is Ralph L. Andreano's Ph.D. dissertation, "The Emergence of New Competition in the American Petroleum Industry Before 1911." Andreano convincingly presents the thesis that the Standard Oil monopoly had ceased to exist almost a decade before the United States Supreme Court ordered its dissolution in 1911. The discovery of the California and Gulf Coast fields brought a glut of flush production that even Standard Oil could not control. This allowed new oil companies, including Cullinan's Texas Company, to gain a foothold in the American petroleum industry and to challenge the Standard Oil monopoly. Andreano's thesis has been incorporated into a recent book, *The American Petroleum Industry: The Age of Energy, 1899–1959* (1963), of which he is co-author with Harold F. Williamson and others.

Other unpublished manuscripts used in this study appear in the alphabetical listing below.

Public Documents

Federal documents, such as the publications of the United States Geological Survey, the Bureau of the Census, the Bureau of Corporations, and the Industrial Commission, were particularly helpful in tracing the background and development of the Corsicana, Texas, oil field from 1894 through 1901. Indeed, the dearth of manuscript collections concerning this field makes these publications a principal source for investigation of the first substantial petroleum development in Texas. Documents used in this work are listed below.

Books and Essays

Several books merit special attention because of their extensive use in this work. Charles A. Warner's *Texas Oil and Gas Since 1543*, with its detailed chronology and geological discussions, is the standard study of the state's petroleum industry. It is also very valuable because of its statistics and tables, which give petroleum production for separate fields, counties, and geographical areas of Texas. It is unfortunate, however, that this study has not been revised and extended to cover Texas petroleum development in the decades following World War II.

This need is partially met by Carl C. Rister's *Oil! Titan of the Southwest*, which deals with southwestern petroleum development by states through the 1940s. Students of the pre-Spindletop petroleum industry in Texas will be disappointed, however. Rister closely follows Warner's earlier work and adds little, for instance, to the history of the early Corsicana field's development.

James A. Clark and Michel T. Halbouty's *Spindletop* is an important work which captures the drama and impact of the birth of the Gulf Coast petroleum industry. Although written in a popular vein, without specific documentation and footnotes, it is nevertheless a reliable work based upon extensive research by its authors—one a long-time journalist-historian of the petroleum industry and the other a respected geologist.

The classic study of America's dominant oil company by Ralph W. and Muriel E. Hidy, *History of Standard Oil Company (New Jersey): Pioneering in Big Business, 1882–1911,* was very helpful in furnishing background for Cullinan's earlier career, the Corsicana field development, and for tracing biographical details of a few former employees of Standard who joined Cullinan in Texas.

Another business history furnished important background information on the legal climate in Texas during the early twentieth-century petroleum development: Henrietta M. Larson and Kenneth Wiggins Porter's *History of Humble Oil and Refining Company: A Study in Industrial Growth.*

Two works which attempt to survey major trends in the development of the entire American petroleum industry were also consulted: Harold F. Williamson and Arnold R. Daum, *The American Petroleum Industry: The Age of Illumination, 1859–1899;* and Harold F. Williamson, Ralph L. Andreano, Arnold R. Daum, and Gilbert C. Klose, *The American Petroleum Industry: The Age of Energy, 1899–1959.* These books were helpful in reviewing technological innovation within the industry as well as furnishing tables and charts concerning pricing and production statistics. These and other books and essays used are listed below.

Periodicals

The files of the Dallas *Morning News* from 1894 through 1901 are essential to research on the Corsicana oil field. Because of Corsicana's proximity to Dallas (sixty miles) the paper could readily cover the petroleum activity in that area, and it took particular interest in the role of J. S. Cullinan in the development of that field.

Among trade papers and articles, the files of the *Oil Investors' Journal,* 1902 through 1910, are an excellent source for the development of the entire Gulf Coast petroleum area, beginning with the Spindletop discovery. Its successor publication, the *Oil and Gas Journal,* 1910 to date, was also used extensively in this work, particularly for the years 1913 through 1916.

Other newspaper files, trade papers, and articles used in this study appear below.

Andreano, Ralph L. "The Emergence of New Competition in the American Petroleum Industry Before 1911." Ph.D. dissertation, Northwestern University, 1960.

Austin, Texas. Memorial Library, University of Texas. Louis Wiltz Kemp papers.

Ball, Max W. *This Fascinating Oil Business.* New York: Bobbs-Merrill Company, 1940.

Brantly, J. E. "Hydraulic Rotary-Drilling System," in *History of Petroleum Engineering,* edited by D. V. Carter. Dallas: American Petroleum Institute, 1961.

———. "Percussion-Drilling System," in *History of Petroleum Engineering,* edited by D. V. Carter. Dallas: American Petroleum Institute, 1961.

Carothers, Durrell. "Roger Q. Mills." Paper presented at 1930 seminar, Rice Institute, Houston.

Chicago *Daily Trade Bulletin.* October 31, 1903.

Clark, James A. *The Chronological History of the Petroleum and Natural Gas Industries.* Houston: Clark Book Company, 1963.

Clark, James A., and Michel T. Halbouty. *Spindletop.* New York: Random House, 1952.

Coberly, C. J. "Production Equipment," in *History of Petroleum Engineering,* edited by D. V. Carter. Dallas: American Petroleum Institute, 1961.

Cotner, Robert C. *James Stephen Hogg: A Biography.* Austin: University of Texas Press, 1959.

Dallas *Morning News.* 1894–1901.

Davis, Ellis A., and Edwin H. Grobe, editors and compilers. *The New Encyclopedia of Texas,* vols. I and II. Dallas: Texas Development Bureau, n.d. [ca. 1925].

Derrick's Hand-Book of Petroleum, vols. I and II. Oil City, Pennsylvania: Derrick Publishing Company, 1898–1900.

Drucker, Peter F. *The Concept of the Corporation.* New York: New American Library, Mentor edition, 1964.

Finty, Tom, Jr. *Anti-Trust Legislation in Texas: An Historical and Analytical Review of the Enactment and Administration of the Various Laws Upon the Subject.* Dallas: A. H. Bello & Company, 1916.

Galveston *News.* 1903.

Harris, Innis D. "Legal History of Conservation of Oil and Gas in Kansas," in *Legal History of Conservation of Oil and Gas: A Symposium.* Chicago: American Bar Association, 1939.

Hidy, Ralph W., and Muriel E. Hidy. *History of the Standard Oil Company (New Jersey): Pioneering in Big Business, 1882–1911.* New York: Harper and Brothers, 1955.

Houston, Texas. M. D. Anderson Library, University of Houston. Texas Gulf Coast Historical Association Archives. Joseph Stephen Cullinan papers.

Houston, Texas. M. D. Anderson Library, University of Houston. Texas Gulf Coast Historical Association Archives. The Charles A. Warner Collection.

Houston, Texas. W. W. Fondren Library, William Marsh Rice University. James Lockhart Autry papers.

Houston *Chronicle.* 1913, 1937, 1970.

Houston *Post.* 1905, 1934–1937, 1968.

Houston *Press.* 1922, 1937.

James, Marquis. *The Texas Story: The First Fifty Years.* New York: The Texas Company, 1953.

Josephson, Matthew. *The Robber Barons: The Great American Capitalists, 1861–1901.* Revised edition. New York: Harcourt, Brace & World, Inc., 1962.

Kemp, Louis W., and Wilfred B. Talman, compilers. "Documentary History of the Texas Company." New York: The Texas Company, n.d. [ca. 1950].

King, John O. *The Early History of the Houston Oil Company of Texas, 1901–1908.* Houston: Texas Gulf Coast Historical Association, 1959.

Larson, Henrietta M., and Kenneth Wiggins Porter. *History of Humble Oil and Refining Company: A Study in Industrial Growth.* New York: Harper and Brothers, 1959.

Magnolia Oil News. Founders' Number, April 1931.

Mobil Oil Company Publications Staff. "History of the Refining Department of the Magnolia Petroleum Company." Beaumont, Texas: Mobil Oil Company, n.d.

National Oil Reporter. March 28, 1903.

New York. The Texas Company, Chrysler Building. The Texas Company Archives.

O'Connor, Harvey. *Mellon's Millions.* New York: John Day Company, 1933.

Oil Investors' Journal. 1902–1910.

Oil and Gas Journal. 1910–1969.

Port Arthur, Texas. W.P.A. American Guide Series. Houston: Anson Jones Press, 1940.

Randall, Clarence B. *The Folklore of Management.* New York: New American Library, Mentor edition, 1962.

Reed, S. G. *History of Texas Railroads.* Houston: St. Clair Publishing Company, 1941.

Rister, Carl C. *Oil! Titan of the Southwest.* Norman: University of Oklahoma Press, 1949.

Summers, Walter L. "The Modern Theory and Practical Application of Statutes for the Conservation of Oil and Gas," in *Legal History of Conservation of Oil and Gas: A Symposium.* Chicago: American Bar Association, 1939.

Taylor, Ava. *Navarro County: History and Photographs.* Corsicana, Texas: Taylor Publishing Company, 1962.

The Texaco Star. XIV (February 1927), 33.

Texas. *Annual Report of the State Railroad Commission, 1897.* Austin: Steckman Printers, 1898.

———. Office of the Attorney General. *Report of the Attorney-General, For the Years 1897–98: M. M. Crane.* Austin: Von Boeckmann, Moore and Schutze, State Contractors, 1899.

———. Office of the Attorney General. *Report and Opinions of the Attorney General, For the Years 1908–1910: Jewel P. Lightfoot.* Austin: Austin Printing Co., Printers, 1911.

———. Office of the Attorney General. *Biennial Report of the Attorney General, 1912–1914: B. F. Looney.* Austin: Von Boeckmann-Jones Company, Printers, 1915.

———. *General Laws of the State of Texas, Twenty-Ninth Legislature, Regular Session, 1905.* Austin: State Printing Company, 1905.

———. *House Journal, Twenty-Sixth Legislature, 1899.* Austin: State Printers, 1900.

———. *House Journal, Twenty-Ninth Legislature, 1905.* Austin: State Printers, 1905.

———. Jefferson County (Beaumont) Deed Records, vol. LIII.

———. *Journal of the Senate, Twenty-Sixth Legislature, 1899.* Austin: State Printers, 1900.

———. *Laws of Texas, 1897–1902,* compiled by H. P. N. Gammel. Austin: Gammel Book Company, 1902.

———. Navarro County (Corsicana) Deed Records, vol. XCII.

———. Secretary of State John G. Tod. *Principal Corporation Laws of the State of Texas.* Austin: Von Boeckmann Printers, 1902.

———. Secretary of State John G. McKay. *Principal Corporation Laws of the State to Date.* Austin: A. C. Baldwin and Sons, 1916.

———. *Texas Petroleum,* by William B. Phillips. Mineral Survey Bulletin No. 1. Austin: University of Texas, 1901.

Thompson, Craig. *Since Spindletop: A Human Story of Gulf's First Half-Century.* Pittsburgh: Gulf Oil Corporation, 1951.

U.S. Bureau of the Census. *Historical Statistics of the United States, 1790–1957.* Washington: Government Printing Office, 1960.

———. Bureau of the Census. *Tenth Census of the United States: 1880. Cotton Production in the United States, I.* Washington: Government Printing Office, 1884.

———. Bureau of the Census. *Eleventh Census of the United States: 1890. Population, I. Manufacturing, II.* Washington: Government Printing Office, 1893.

———. Bureau of the Census. *Twelfth Census of the United States: 1900. Population, I. Agriculture, II.* Washington: Government Printing Office, 1901, 1902.

———. Bureau of Corporations. *Report of the Commissioner of Corporations on the Petroleum Industry: Part I, Position of the Standard Oil Company in the Petroleum Industry; Part II, Prices and Profits.* Washington: Government Printing Office, 1907.

———. Department of the Interior. *The Corsicana Oil and Gas Field, Texas,* by George C. Matson and Oliver B. Hopkins. U.S. Geological Survey Bulletin No. 661F. Washington: Government Printing Office, 1918.

———. Department of the Interior. *Mineral Resources of the United States, Nonmetals. 1889–1900*, II, III. U.S. Geological Survey. Washington: Government Printing Office, 1890–1901.

———. Department of the Interior. *Oil and Gas Fields of the Western Interior and Northern Texas Coal Measures and of the Upper Cretaceous and Tertiary of the Western Gulf Coast*, by George I. Adams. U.S. Geological Survey Bulletin No. 184. Washington: Government Printing Office, 1901.

———. Department of the Interior. *The Southwestern Coal Field*, by Joseph A. Taff. U.S. Geological Survey, 22nd Annual Report, III. Washington: Government Printing Office, 1902.

———. Industrial Commission. *Report of the Industrial Commission on Trusts and Industrial Companies*, XIII, pt. 2. Washington: Government Printing Office, 1901.

Warner, Charles A. *Texas Oil and Gas Since 1543*. Houston: Gulf Coast Publishing Company, 1939.

Wendt, Lloyd, and Herman Kogan. *Bet a Million! The Story of John W Gates*. Indianapolis: Bobbs-Merrill Company, 1948.

White, Gerald T. *Formative Years in the Far West, A History of Standard Oil Company of California and Predecessors Through 1919*. New York: Appleton-Century-Crofts, 1962.

Williamson, Harold F., and Arnold R. Daum. *The American Petroleum Industry: The Age of Illumination, 1859–1899*. Evanston: Northwestern University Press, 1959.

Williamson, Harold F.; Ralph L. Andreano; Arnold R. Daum; and Gilbert C. Klose. *The American Petroleum Industry: The Age of Energy, 1899–1959*. Evanston: Northwestern University Press, 1963.

Court Cases Cited

Beattie v. Hardy, 53 Southwest Reports 685 (Texas, 1899).

Cullinan v. Walker, as Collector of Internal Revenue for the First District of Texas, 262 U.S. 134 (1923).

Ramsey et al. v. Tod, 69 Southwest Reports 133 (Texas, 1902).

State v. E. T. Hathaway, 36 Texas Criminal Reports 261 (1896).

Index

Akin, Emlin, 3, 21
Allyn, Charles H., 13, 50
American Petroleum Company, 199–200
American Republics Corporation: organization of, 200–201; subsidiaries of, 201–202; oil properties of, 208–209; proxy battle, 209–210; liquidation of, 210
American Steel and Wire Company, 105
American Well and Prospecting Company, 13, 21
Andreano, Ralph L., 94
Arriola, Texas, oil field, 206
August Belmont and Company, 189
Autry, James L.: Corsicana activities, 12, 12–13, 17; advises J. S. Cullinan, 56, 121, 122, 126, 129–131, 168; biographical sketch of, 77; Texas Company officer, 135, 153, 168, 190; resigns from Texas Company, 193; Farmers Petroleum Company stockholder, 198

Bailey, Joseph W., 124, 125
Baker, C. E., 21
Baker, M. C., 21
Barrett, Lyne T., 9
Beaton, Ralph, 14, 17, 62
Beattie v. *Hardy*, 131
Beaty, Amos L., 156–157, 193
Bethlehem Steel Corporation, 22
Boyle, Patrick C., 95
Brooks, R. E.: oil interest of, 97, 101, 119; Texas Company director, 135, 143–144, 187
Brown, Edwy R., 53
Bryan bill, 171, 174–175
Bryan, Chester H., 171
Buckeye Pipe Line Company, 5
Buckner, Eric H., 78
Burt, Andrew D., 197
Burt Refining Company, 123

Caddo, Louisiana, oil field, 184
Calhoun, Patrick, 66, 203, 205
Campbell, William T.: oil interests of,

98, 101, 112, 115, 119; banking interest of, 108; proposed sale to British investors, 134; Texas Company director, 135, 147
Carroll, Ernest, 78, 153
Carroll, George W., 85
Carroll, Guy, 77–78, 153
Central Oil Refining Company, 79
Citizens National Bank of Port Arthur, Texas, 108
Coal prices, Texas, 24–25
Colligan, John C., 154
Consumers Oil Company, 17
Corsicana, Texas: agricultural center, 11–12; streets treated with oil, 44; oil boom impact upon, 72–74, 76–77, 78; population growth of, 74–75; local residents in oil industry, 77–78
Corsicana, Texas, oil field: oil discovered, 13–14; production of, 18–19, 79; geological characteristics, 20; drilling practices, 19–22, 39–40; storage, 22–23, 26–27, 40; refining, 23–24, 80–81; prices, 24–26, 40–42; natural gas, 27–28, 44–46; wages, 74; canteens established, 75; pipelines, 80
Corsicana Commercial Club, 12
Corsicana Gas and Electric Light Company, 44–45
Corsicana Oil Development Company, 14–15, 17, 28
Corsicana Petroleum Company: leases acquired by, 57, 59–61; organization of, 58–59; production of, 61; profitability of, 61; acquired by Magnolia Petroleum Company, 61; Cullinan sells interest, 61
Corsicana Refining Company: equipment of, 48–49; products of, 49, 50–51; marketing of, 49–50; output of, 51; acquired by Magnolia Petroleum Company, 80–81
Corsicana Water Development Company, 12, 56, 57